Polynomial Functional
Dynamical Systems

大江東去浪淘盡

千古風流事

滿腔寬屈空誰朝傻

Synthesis Lectures on Mechanical Engineering

Synthesis Lectures on Mechanical Engineering series publishes 60–150 page publications pertaining to this diverse discipline of mechanical engineering. The series presents Lectures written for an audience of researchers, industry engineers, undergraduate and graduate students.

Additional Synthesis series will be developed covering key areas within mechanical engineering.

The Engineering Dynamics Course Companion, Part 2: Rigid Bodies: Kinematics and Kinetics
Edward Diehl
2020

The Engineering Dynamics Course Companion, Part 1: Particles: Kinematics and Kinetics
Edward Diehl
2020

Fluid Mechanics Experiments
Robabeh Jazaei
2020

Sequential Bifurcation Trees to Chaos in Nonlinear Time-Delay Systems
Siyuan Xing and Albert C.J. Luo
2020

Introduction to Deep Learning for Engineers: Using Python and Google Cloud Platform
Tariq M. Arif
2020

Towards Analytical Chaotic Evolutions in Brusselators
Albert C.J. Luo and Siyu Guo
2020

Modeling and Simulation of Nanofluid Flow Problems
Snehashish Chakraverty and Uddhaba Biswal
2020

Modeling and Simulation of Mechatronic Systems using Simscape
Shuvra Das
2020

Automatic Flight Control Systems
Mohammad Sadraey
2020

Bifurcation Dynamics of a Damped Parametric Pendulum
Yu Guo and Albert C.J. Luo
2019

Reliability-Based Mechanical Design, Volume 2: Component under Cyclic Load and Dimension Design with Required Reliability
Xiaobin Le
2019

Polynomial Functional Dynamical Systems

Albert C. J. Luo

ISBN: 978-3-031-79708-8   paperback
ISBN: 978-3-031-79709-5   ebook
ISBN: 978-3-031-79710-1   hardcover

DOI 10.1007/978-3-031-79709-5

A Publication in the Springer series
SYNTHESIS LECTURES ON MECHANICAL ENGINEERING

Lecture #38
Series ISSN
Print 2573-3168   Electronic 2573-3176

Polynomial Functional Dynamical Systems

Albert C. J. Luo
Southern Illinois University, Edwardsville

SYNTHESIS LECTURES ON MECHANICAL ENGINEERING #38

ABSTRACT

The book is about the global stability and bifurcation of equilibriums in polynomial functional systems. Appearing and switching bifurcations of simple and higher-order equilibriums in the polynomial functional systems are discussed, and such bifurcations of equilibriums are not only for simple equilibriums but for higher-order equilibriums. The third-order sink and source bifurcations for simple equilibriums are presented in the polynomial functional systems. The third-order sink and source switching bifurcations for saddle and nodes are also presented, and the fourth-order upper-saddle and lower-saddle switching and appearing bifurcations are presented for two second-order upper-saddles and two second-order lower-saddles, respectively. In general, the $(2l + 1)$th-order sink and source switching bifurcations for $(2l_i)$th-order saddles and $(2l_j + 1)$-order nodes are also presented, and the $(2l)$th-order upper-saddle and lower-saddle switching and appearing bifurcations are presented for $(2l_i)$th-order upper-saddles and $(2l_j)$th-order lower-saddles $(i, j = 1, 2, \ldots)$. The vector fields in nonlinear dynamical systems are polynomial functional. Complex dynamical systems can be constructed with polynomial algebraic structures, and the corresponding singularity and motion complexity can be easily determined.

KEYWORDS

functional dynamical systems, appearing bifurcations, switching bifurcations, sink bifurcations, source bifurcations, upper-saddle bifurcations, lower-saddle bifurcations

Contents

Preface

In this book, the global stability and bifurcation of equilibriums in low-degree polynomial functional systems are presented. Appearing and switching bifurcations of simple and higher-order equilibriums in the polynomial functional systems are discussed, and such bifurcations of equilibriums are not only for simple equilibriums but for higher-order equilibriums. The third-order sink and source bifurcations for simple equilibriums are presented in the polynomial functional systems. The third-order sink and source switching bifurcations for *saddle* and nodes are also presented, and the fourth-order *upper-saddle* and *lower-saddle* switching and appearing bifurcations are obtained for two second-order *upper-saddles* and two second-order *lower-saddles*, respectively.

This book has six chapters. Chapter 1 discusses linear functional systems. Chapter 2 discussed quadratic functional dynamical systems. The appearing and switching bifurcations of functional equilibriums are presented with and without functional singularity. In Chapter 3, the singularity and bifurcations for cubic functional dynamical systems are discussed. Chapter 4 discusses quartic functional systems. The higher-order functional equilibriums and switching bifurcations are presented. Chapter 5 extended the singularity of bifurcations in quartic functional systems to the $(2m)$th-order polynomial functional dynamical system. The appearing and switching functional bifurcations are presented. In Chapter 6, based on the cubic and quartic functional dynamical systems, the appearing and switching functional bifurcations in the $(2m + 1)$th polynomial functional dynamical systems are presented.

Finally, the author hopes the materials presented herein can help us construct more complex dynamical systems based on polynomial algebraic structures, and the corresponding singularity and motion complexity can be easily determined.

Albert C. J. Luo
Edwardsville, Illinois
August 2021

CHAPTER 1

Linear Functional Systems

In this chapter, the stability and stability switching of equilibriums in linear functional systems are discussed. The sink and source equilibriums in the linear functional systems are discussed.

Definition 1.1 Consider a 1-dimensional linear functional dynamical system

$$\dot{x} = A(\mathbf{p})g(x) + B(\mathbf{p}), \tag{1.1}$$

where two scalar constants $A(\mathbf{p})$ and $B(\mathbf{p})$ are determined by a vector parameter

$$\mathbf{p} = (p_1, p_2, \ldots, p_m)^{\mathrm{T}}. \tag{1.2}$$

(i) If $A(\mathbf{p}) \neq 0$, there is an equilibrium set S_1 as

$$x^* \in S_1 = \{a_1^{(i)} | g(a_1^{(i)}) = a_1, \ i = 1, 2, \ldots, N\} \tag{1.3}$$

for a functional equilibrium of

$$g(x^*) = a_1(\mathbf{p}) = -\frac{B(\mathbf{p})}{A(\mathbf{p})}, \ \text{with } a_0(\mathbf{p}) = A(\mathbf{p}). \tag{1.4}$$

(i$_1$) If $x^* \in S_1 \neq \{\emptyset\}$, the corresponding functional dynamical system with equilibrium x^* becomes

$$\dot{x} = a_0[g(x) - a_1]. \tag{1.5}$$

(i$_2$) If $x^* \in S_1 = \{\emptyset\}$, the flow of the functions dynamical system in Eq. (1.1) is

- positive for $a_0[g(x) - a_1] > 0$;
- negative for $a_0[g(x) - a_1] < 0$.

(ii) If $A(\mathbf{p}) = 0$, Eq. (1.5) becomes

$$\dot{x} = B(\mathbf{p}). \tag{1.6}$$

For $B(\mathbf{p}) \neq 0$, the 1-dimensional functional system is called a constant velocity system.

For $B(\mathbf{p}) = 0$, the 1-dimensional functional system is called a permanent static system with zero velocity.

(iii) For $\|\mathbf{p}\| \to \|\mathbf{p}_0\| = \beta$, if the following relations hold

$$A(\mathbf{p}) = a_0 = \varepsilon \to 0, \quad B(\mathbf{p}) = \varepsilon a_1(\mathbf{p}) \to 0, \tag{1.7}$$

then there is an instant equilibrium to the vector parameter $\mathbf{p}$ determined by

$$g(x^*) = a_1(\mathbf{p}). \tag{1.8}$$

with $x^* \in S_1 \neq \{\emptyset\}$.

Theorem 1.2   *Under assumption* (1.7), *a standard form of the 1-dimensional dynamical system in Eq.* (1.1) *is*

$$\dot{x} = f(x) = a_0[g(x) - a_1] \tag{1.9}$$

with $x^* \in S_1 \neq \{\emptyset\}$.

(i) *If $df/dx|_{x^*=a_1^{(i)}} = a_0 dg/dx|_{x^*=a_1^{(i)}} < 0$, then the equilibrium of $x^* = a_1^{(i)} \in S_1$ simply with $g(x^*) = a_1(\mathbf{p})$ is stable. Such a stable equilibrium is called a sink or a stable node.*

(ii) *If $df/dx|_{x^*=a_1^{(i)}} = a_0 dg/dx|_{x^*=a_1^{(i)}} < 0$, then equilibrium of $x^* = a_1^{(i)} \in S_1$ simply with $g(x^*) = a_1(\mathbf{p})$ is unstable. Such an unstable equilibrium is called a source or an unstable node.*

(iii) *If $df/dx|_{x^*=a_1^{(i)}} = a_0 dg/dx|_{x^*=a_1^{(i)}} = 0$ but $a_0 \neq 0$, then equilibrium of $x^* = a_1^{(i)} \in S_1$ simply with $g(x^*) = a_1(\mathbf{p})$ is critical. Such a critical equilibrium is called a singular node.*

(iv) *If $a_0(\mathbf{p}) = 0$, then the flow in a neighborhood of functional equilibrium of $x^* = a_1^{(i)} \in S_1$ simply with $g(x^*) = a_1(\mathbf{p})$ is static (critical). Such a static functional equilibrium is called a functional critical case.*

Proof. For $x^* \in S_1 \neq \{\emptyset\}$, let $y = x - a_1^{(i)}$ and $\dot{x} = \dot{y}$. Thus, Eq. (1.9) becomes

$$\dot{y} = df/dx|_{x^*=a_1^{(i)}} y = (a_0 dg/dx|_{x^*=a_1^{(i)}})y = \lambda y$$

$$\text{for } a_1^{(i)} \in S_1, \ \lambda = a_0 dg/dx|_{x^*=a_1^{(i)}}.$$

The corresponding solution is

$$y = y_0 e^{\lambda(t-t_0)},$$

where $y_0 = y(t_0) = x_0 - a_1^{(i)}$ is an initial condition.

(i) If $\lambda < 0$, we have

$$\lim_{t\to\infty} (x - a_1^{(1)}) = \lim_{t\to\infty} y = \lim_{t\to\infty} y_0 e^{\lambda(t-t_0)} = 0 \;\Rightarrow\; \lim_{t\to\infty} x(t) = a_1^{(i)}.$$

So the equilibrium of $x^* = a_1^{(i)}(\mathbf{p})$ simply with $g(x^*) = a_1$ is stable.

(ii) If $\lambda > 0$, we have

$$\lim_{t\to\infty} (x - a_1) = \lim_{t\to\infty} y = \lim_{t\to\infty} y_0 e^{\lambda(t-t_0)} = \infty \;\Rightarrow\; \lim_{t\to\infty} x(t) = \infty.$$

So the equilibrium of $x^* = a_1^{(i)}(\mathbf{p})$ simply with $g(x^*) = a_1$ is unstable.

(iii) If $\lambda = 0$ but $a_0 \neq 0$, we have

$$\lim_{t\to\infty} (x - a_1^{(i)}) = \lim_{t\to\infty} y = \lim_{t\to\infty} y_0 e^{\lambda(t-t_0)} = y_0.$$

Thus, the higher-order Taylor series should be considered for the higher singularity. Such an equilibrium is called a higher-order singular equilibrium.

(iv) If $\lambda = 0$ with $a_0 = 0$, we have

$$\lim_{t\to\infty} (x - a_1^{(i)}) = \lim_{t\to\infty} y = \lim_{t\to\infty} y_0 e^{\lambda(t-t_0)} = y_0 \;\Rightarrow\; \lim_{t\to\infty} x(t) = x_0.$$

So the flow in the neighborhood of equilibrium $x^* = a_1^{(i)}(\mathbf{p})$ is static.
The theorem is proved. □

For function $g(x)$ in the linear functional systems, the following cases exist.

- If $g(x)$ is a linear function, such a linear functional system is a linear dynamical system. The stability of equilibrium can be determined through the eigenvalue analysis.

- If $g(x)$ is a nonlinear function, such a linear functional system is a nonlinear dynamical system, and the corresponding local stability and bifurcations of equilibriums can be determined as in Luo (2020).[1]

- If $g(x)$ is a polynomial function, such a linear functional system is a polynomial nonlinear dynamical system, and the corresponding stability and bifurcations of equilibrium were presented in Luo (2020).[1]

Similarly, the function $g(x)$ can be defined by a generalized m-time functional function as

$$g(x) = g_1(g_2(\dots(g_m(x)))). \tag{1.10}$$

---

[1] Albert C. J. Luo (2020), *Bifurcation and Stability in Nonlinear Dynamical Systems*, Springer, Switzerland.

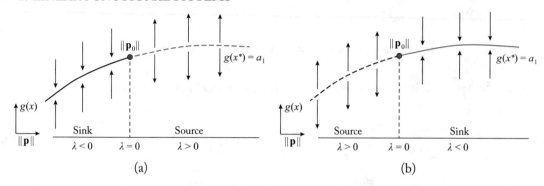

Figure 1.1: Stability of single functional equilibrium in the 1-dimensional linear functional dynamical system: (a) left stable equilibrium ($\lambda = a_0 dg/dx|_{x^*} < 0$) and (b) right stable equilibrium ($\lambda = a_0 dg/dx|_{x^*} < 0$). Stable and unstable equilibriums are represented by solid and dashed curves, respectively. The stability switching is labeled by a circular symbol.

To illustrate the stability of equilibrium, one equilibrium point of $x^* = a_1^{(i)}(\mathbf{p})$ simply with $g(x^*) = a_1$ changes with a vector parameter $\mathbf{p}$. The stability of such an equilibrium is determined by $\lambda = a_0 dg/dx|_{x^*=a_1^{(i)}}$. The stability switching is at the boundary $\mathbf{p}_0 \in \partial\Omega_{12}$ with $a_0 = 0$ if $dg/dx|_{x^*=a_1^{(i)}} \neq 0$. The stable functional equilibrium on the left and right sides are presented on Figs. 1.1a,b, respectively. The stable and unstable portions of the equilibrium are presented by the solid and dash curves, respectively.

CHAPTER 2

Quadratic Nonlinear Functional Systems

In this chapter, the stability of equilibriums in 1-dimensional quadratic nonlinear functional systems are discussed. The *upper-saddle-node* and *lower-saddle-node* appearing and switching bifurcations with and without functional singularity are discussed.

2.1 QUADRATIC FUNCTIONAL SYSTEMS

In this section, a general theory for quadratic functional system will be presented and the appearing and switching bifurcation will be discussed.

2.1.1 A GENERAL THEORY

Definition 2.1 Consider a 1-dimensional quadratic nonlinear functional dynamical system as

$$\dot{x} = A(\mathbf{p})(g(x))^2 + B(\mathbf{p})g(x) + C(\mathbf{p}), \tag{2.1}$$

where three scalar constants $A(\mathbf{p}) \neq 0$, $B(\mathbf{p})$, and $C(\mathbf{p})$ are determined by a vector parameter

$$\mathbf{p} = (p_1, p_2, \ldots, p_m)^{\mathrm{T}}. \tag{2.2}$$

(i) If

$$\Delta = B^2 - 4AC < 0 \text{ for } \mathbf{p} \in \Omega_1 \subset \mathbf{R}^m, \tag{2.3}$$

then the quadratic nonlinear functional system does not have any functional equilibriums. The flow without equilibrium is called a non-equilibrium flow.

(i_1) If $a_0(\mathbf{p}) = A(\mathbf{p}) > 0$, the non-equilibrium flow is called a positive flow.

(i_2) If $a_0(\mathbf{p}) = A(\mathbf{p}) < 0$, the non-equilibrium flow is called a negative flow.

(ii) If

$$\Delta = B^2 - 4AC > 0 \text{ for } \mathbf{p} \in \Omega_2 \subset \mathbf{R}^m, \tag{2.4}$$

then the quadratic nonlinear functional system has two different sets of equilibriums as

$$\begin{aligned}
x^* \in S_1 &= \{a_1^{(i)} | g(a_1^{(i)}) = a_1, \ i = 1, 2, \ldots, N_1\} \cup \{\emptyset\} \\
x^* \in S_2 &= \{a_2^{(i)} | g(a_2^{(i)}) = a_2, \ i = 1, 2, \ldots, N_2\} \cup \{\emptyset\}
\end{aligned} \tag{2.5}$$

with

$$g(x^*) = a_1 \text{ and } g(x^*) = a_2, \tag{2.6}$$

and the corresponding standard functional form is

$$x^* = a_0(g(x) - a_1)(g(x) - a_2), \tag{2.7}$$

where

$$a_0 = A(\mathbf{p}), \ a_{1,2} = \frac{-B(\mathbf{p}) \pm \sqrt{\Delta}}{2A(\mathbf{p})} \text{ with } a_1 < a_2. \tag{2.8}$$

(iii) If

$$\Delta = B^2 - 4AC = 0 \text{ for } \mathbf{p} = \mathbf{p}_0 \in \partial\Omega_{12} \subset \mathbf{R}^{m-1}, \tag{2.9}$$

then the 1-dimensional functional dynamical system has a set of double-repeated equilibriums, i.e.,

$$x^* \in S_1 = \{a_1^{(i)} | g(a_1^{(i)}) = a_1, \ i = 1, 2, \ldots, N_1\} \tag{2.10}$$

with

$$g(x^*) = a_1 \text{ and } g(x^*) = a_1, \tag{2.11}$$

with the corresponding standard functional form of

$$x^* = a_0(g(x) - a_1)^2, \tag{2.12}$$

where

$$a_0 = A(\mathbf{p}_0), \text{ and } a_1 = a_2 = -\frac{B(\mathbf{p}_0)}{2A(\mathbf{p}_0)}. \tag{2.13}$$

Such a flow with the equilibrium of $x^* \in S_1$ simply with $g(x^*) = a_1(\mathbf{p})$ is called a *saddle* flow of the second order.

(iii$_1$) If $a_0(\mathbf{p}) > 0$, then the equilibrium of $x^* \in S_1$ simply with $g(x^*) = a_1(\mathbf{p})$ is an *upper-saddle*.

(iii$_2$) If $a_0(\mathbf{p}) < 0$, then the equilibrium of $x^* \in S_1$ simply with $g(x^*) = a_1(\mathbf{p})$ is a *lower-saddle*.

(iv) The equilibrium of $x^* \in S_1$ simply with $g(x^*) = a_1(\mathbf{p})$ for two simple equilibriums vanishing or appearance is called a *saddle-node* bifurcation of equilibrium at a point $\mathbf{p} = \mathbf{p}_0 \in \partial\Omega_{12}$, and the bifurcation condition is

$$\Delta = B^2 - 4AC = 0. \tag{2.14}$$

(iv$_1$) If $a_0(\mathbf{p}) > 0$, the bifurcation at $x^* \in S_1$ simply with $g(x^*) = a_1(\mathbf{p})$ for two simple equilibriums appearance or vanishing is called an *upper-saddle-node* bifurcation.

(iv$_2$) If $a_0(\mathbf{p}) < 0$, the bifurcation at $x^* \in S_1$ simply with $g(x^*) = a_1(\mathbf{p})$ for two simple equilibriums appearance or vanishing is called a *lower-saddle-node* bifurcation.

Theorem 2.2

(i) *Under a condition of*

$$\Delta = B^2 - 4AC < 0, \tag{2.15}$$

a standard functional form of the 1-dimensional dynamical functional system in Eq. (2.1) *is*

$$x^* = a_0[(g(x) - \frac{1}{2}\frac{B}{A})^2 + \frac{1}{4A^2}(-\Delta)] \tag{2.16}$$

with $a_0 = A(\mathbf{p})$, *which has a non-equilibrium flow.*

(i$_1$) *If* $a_0(\mathbf{p}) > 0$, *the non-equilibrium flow is called a positive flow.*

(i$_2$) *If* $a_0(\mathbf{p}) > 0$, *the non-equilibrium flow is called a negative flow.*

(ii) *Under a condition of*

$$\Delta = B^2 - 4AC > 0, \tag{2.17}$$

a standard functional form of the quadratic nonlinear functional dynamical system in Eq. (2.1) *is*

$$\dot{x} = f(x, \mathbf{p}) = a_0(g(x) - a_1)(g(x) - a_2). \tag{2.18}$$

(ii$_1$) *The simple equilibrium of* $x^* = a_1^{(i)} \in S_1$ *simply with* $g(x^*) = a_1(\mathbf{p})$ *is stable (sink) with* $df/dx|_{x^*=a_1^{(i)}} < 0$, *or unstable (source) with* $df/dx|_{x^*=a_1^{(i)}} > 0$.

(ii$_2$) *The simple equilibrium of* $x^* = a_2^{(i)} \in S_2$ *simply with* $g(x^*) = a_2(\mathbf{p})$ *is stable (sink) with* $df/dx|_{x^*=a_2^{(i)}} < 0$ *or unstable (source) with* $df/dx|_{x^*=a_2^{(i)}} > 0$.

(iii) *Under a condition of*

$$\Delta = B^2 - 4AC = 0, \text{ and } dg/dx|_{x^* \in S_i} \neq 0 \tag{2.19}$$

a standard functional form of the quadratic functional dynamical system in Eq. (2.1) *is*

$$\dot{x} = f(x, \mathbf{p}) = a_0(g(x) - a_1)^2. \tag{2.20}$$

(iii$_1$) *If* $a_0(\mathbf{p}) > 0$ *and* $dg/dx|_{x^*=a_1^{(i)}} \neq 0$, *the equilibrium of* $x^* = a_1^{(i)} \in S_1$ *simply with* $g(x^*) = a_1(\mathbf{p})$ *is an upper-saddle of the second order with* $d^2 f/dx^2|_{x^*=a_1^{(i)}} > 0$. *The functional bifurcation at* $x^* = a_1^{(i)} \in S_1$ *simply with* $g(x^*) = a_1(\mathbf{p})$ *for two simple equilibriums appearance or vanishing is called an upper-saddle-node bifurcation.*

(iii$_2$) If $a_0(\mathbf{p}) < 0$ and $dg/dx|_{x^*=a_1^{(i)}} \neq 0$, the equilibrium of $x^* = a_1^{(i)} \in S_1$ simply with $g(x^*) = a_1(\mathbf{p})$ is a lower-saddle of the second order with $d^2 f/dx^2|_{x^*=a_1^{(i)}} < 0$. The functional bifurcation at $x^* = a_1^{(i)} \in S_1$ with $g(x^*) = a_1(\mathbf{p})$ for two simple equilibriums appearance or vanishing is called a lower-saddle-node bifurcation.

(iv) Under a condition of

$$\Delta = B^2 - 4AC > 0, \text{ and } dg/dx|_{x^* \in a_j^{(i)}} = 0, \ d^2g/dx^2|_{x^*=a_j^{(i)}} \neq 0 \qquad (2.21)$$

a standard form of the quadratic nonlinear functional dynamical system in Eq. (2.1) is

$$\dot{x} = f(x, \mathbf{p}) = a_0(g(x) - a_1)(g(x) - a_2). \qquad (2.22)$$

(iv$_1$) If $d^2 f/dx^2|_{x^*=a_j^{(i)}} > 0$, the equilibrium of $x^* = a_j^{(i)} \in S_j (j \in \{1, 2\})$ repeatedly with $g(x^*) = a_j(\mathbf{p})$ is an upper-saddle of the second order. The bifurcation at the repeated equilibrium of $x^* = a_j^{(i)} \in S_j$ for the appearance or vanishing of two simple equilibriums of $g(x^*) = a_j(\mathbf{p})$ is called an upper-saddle-node bifurcation.

(iv$_2$) If $d^2 f/dx^2|_{x=a_j^{(i)}} < 0$, the equilibrium of $x^* = a_j^{(i)} \in S_j$ $(j \in \{1, 2\})$ repeatedly with $g(x^*) = a_j(\mathbf{p})$ is a lower-saddle of the second order. The bifurcation at the repeated equilibrium of $x^* = a_j^{(i)} \in S_j$ $(j = 1, 2)$ for the appearance or vanishing of two simple equilibriums of $g(x^*) = a_j(\mathbf{p})$ is called a lower-saddle-node bifurcation.

(v) Under a condition of

$$\Delta = B^2 - 4AC = 0 \text{ and } dg/dx|_{x^* \in S_1} = 0 \qquad (2.23)$$

a standard functional form of the quadratic nonlinear functional dynamical system in Eq. (2.1) is

$$\dot{x} = f(x, \mathbf{p}) = a_0(g(x) - a_1)^2. \qquad (2.24)$$

(v$_1$) If $a_0 > 0$ and $d^2g/dx^2|_{x^*=a_1^{(i)}} \neq 0$, the repeated functional equilibrium of $x^* = a_1^{(i)} \in S_1$ repeatedly with $g(x^*) = a_1(\mathbf{p})$ is an upper-saddle of the fourth order. The functional bifurcation at the repeated equilibrium of $x^* = a_1^{(i)} \in S_1$ for the appearance or vanishing of two second-order upper-saddle equilibriums of $g(x^*) = a_j(\mathbf{p})$ $(j = 1, 2)$ is called a fourth-order upper-saddle-node bifurcation.

(v$_2$) If $a_0 < 0$ and $d^2g/dx^2|_{x^*=a_1^{(i)}} \neq 0$, the repeated functional equilibrium of $x^* = a_1^{(i)} \in S_1$ repeatedly with $g(x^*) = a_1(\mathbf{p})$ is an lower-saddle of the fourth order. The functional bifurcation at the repeated equilibrium of $x^* = a_1^{(i)} \in S_1$ for the appearance or vanishing of two second-order lower-saddle equilibriums of $g(x^*) = a_j(\mathbf{p})$ $(j = 1, 2)$ is called a fourth-order lower-saddle-node bifurcation.

Proof.

(i) Consider

$$\Delta = B^2 - 4AC < 0.$$

(i$_1$) If $a_0 > 0$, we have

$$\dot{x} = a_0[(g(x) + \frac{1}{2}\frac{B}{A})^2 + \frac{1}{4A^2}(-\Delta)] > 0.$$

Thus, such a non-equilibrium flow is called a positive flow.

(i$_2$) If $a_0 < 0$, we have

$$\dot{x} = a_0[(g(x) + \frac{1}{2}\frac{B}{A})^2 + \frac{1}{4A^2}(-\Delta)] < 0.$$

Thus, such a non-equilibrium flow is called a negative flow.

(ii) Let $\Delta x_i = x - a_i$ ($i = 1, 2$) and $\dot{x} = \Delta \dot{x}_i$. Thus, Eq. (2.18) becomes

$$\Delta \dot{x}_i = a_0(a_i - a_j)\frac{dg}{dx}|_{x^*=a_i^{(s)}} \Delta x_i + \frac{1}{2!}\frac{d^2 f}{dx^2}|_{x^*=a_i^{(s)}}\Delta x_i^2$$
$$i, j \in \{1, 2\}; \ j \neq i,$$

where

$$\frac{df}{dx}|_{x^*=a_i^{(s)}} = a_0(a_i - a_1)\frac{dg}{dx}|_{x^*=a_i^{(s)}} + a_0(a_i - a_2)\frac{dg}{dx}|_{x^*=a_i^{(s)}}$$

$$\frac{d^2 f}{dx^2}|_{x^*=a_i^{(s)}} = a_0(a_i - a_1)\frac{d^2 g}{dx^2}|_{x^*=a_i^{(s)}} + 2a_0(\frac{dg}{dx}|_{x^*=a_i^{(s)}})^2$$

$$+ a_0(a_i - a_2)\frac{d^2 g}{dx^2}|_{x^*=a_i^{(s)}}$$

$$(i = 1, 2).$$

Because Δx_i is arbitrary small, we have

$$\Delta \dot{x}_i \approx \lambda_i \Delta x_i \text{ for } \lambda_i \equiv a_0(a_i - a_j)\frac{dg}{dx}|_{x^*=a_i^{(s)}} \neq 0.$$

The corresponding solution is

$$\Delta x_i = \Delta x_{i0}e^{\lambda_i(t-t_0)},$$

where $\Delta x_{i0} = x_0 - a_i$ is an initial condition.

(ii$_1$) If $\lambda_i < 0$, we have

$$\lim_{t\to\infty}(x-a_i) = \lim_{t\to\infty}\Delta x_i = \lim_{t\to\infty}\Delta x_{i0}e^{\lambda_i(t-t_0)} = 0 \;\Rightarrow\; \lim_{t\to\infty}x(t) = a_i.$$

So the equilibrium of $x^* = a_i^{(s)}$ ($s \in \{1, 2, \ldots, N_i\}$) simply with $g(x^*) = a_i(\mathbf{p})$ is stable.

(ii$_{1a}$) For $\lambda_1 = a_0(a_1 - a_2)dg/dx|_{x^*=a_1^{(s)}} < 0$, due to $a_1 - a_2 < 0$, such a simple equilibrium $x^* \in a_1^{(s)} \in S_1$ is stable (sink) for $a_0 dg/dx|_{x^*=a_1^{(s)}} > 0$.

(ii$_{1b}$) For $\lambda_2 = a_0(a_2 - a_1)dg/dx|_{x^*=a_2^{(s)}} < 0$, due to $a_2 - a_1 > 0$, such a simple equilibrium $x^* \in a_2^{(s)} \in S_2$ is stable (sink) for $a_0 dg/dx|_{x^*=a_1^{(s)}} < 0$.

(ii$_2$) If $\lambda_i > 0$, we have

$$\lim_{t\to\infty}(x-a_i) = \lim_{t\to\infty}\Delta x_i = \lim_{t\to\infty}\Delta x_{i0}e^{\lambda_i(t-t_0)} = \infty \;\Rightarrow\; \lim_{t\to\infty}x(t) = \infty.$$

So the equilibrium of $x^* = a_i^{(s)} \in S_i$ simply with $g(x^*) = a_i(\mathbf{p})$ is unstable.

(ii$_{2a}$) For $\lambda_1 = a_0(a_1 - a_2)dg/dx|_{x^*=a_1^{(s)}} > 0$, due to $a_1 - a_2 < 0$, such a simple equilibrium of $x^* = a_1^{(s)} \in S_1$ is unstable (source) for $a_0 dg/dx|_{x^*=a_1^{(s)}} < 0$.

(ii$_{2b}$) For $\lambda_2 = a_0(a_2 - a_1)dg/dx|_{x^*=a_2^{(s)}} > 0$, due to $a_2 - a_1 > 0$, such a simple equilibrium of $x^* = a_2^{(s)} \in S_2$ is unstable (source) for $a_0 dg/dx|_{x^*=a_2^{(s)}} > 0$.

(iii) Under a condition of

$$\Delta = B^2 - 4AC = 0, \text{ and } dg/dx|_{x^*\in S_1} \neq 0$$

we have

$$\dot{x} = f(x, \mathbf{p}) = a_0(g(x) - a_1)^2$$

and

$$a_1 = a_2 = a_i (i = 1, 2)$$

$$\frac{df}{dx}\Big|_{x^*=a_i^{(s)}} = a_0(a_i - a_1)\frac{dg}{dx}\Big|_{x^*=a_i^{(s)}} + a_0(a_i - a_2)\frac{dg}{dx}\Big|_{x^*=a_i^{(s)}} = 0$$

$$\frac{d^2 f}{dx^2}\Big|_{x^*=a_i^{(s)}} = a_0(a_i - a_1)\frac{d^2 g}{dx^2}\Big|_{x^*=a_i^{(s)}} + 2a_0(\frac{dg}{dx}\Big|_{x^*=a_i^{(s)}})^2$$

$$+ a_0(a_i - a_2)\frac{d^2 g}{dx^2}\Big|_{x^*=a_i^{(s)}}$$

$$= 2a_0(\frac{dg}{dx}\Big|_{x^*=a_i^{(s)}})^2.$$

If $a_1(\mathbf{p}) = a_2(\mathbf{p})$, we have

$$\Delta \dot{x}_i = a_0 (dg/dx|_{x^*=a_j^{(s)}})^2 \Delta x_i^2 (s, j \in \{1, 2\}).$$

(iii$_1$) For $a_0 > 0$, $\Delta \dot{x}_i > 0$ exists. So a flow of x reaches to $x^* = a_1^{(i)} \in S_1$ with $g(x^*) = a_1$ from the initial point of $x_0 < a_1^{(i)}$ and it goes to the positive infinity from $x_0 > a_1^{(i)}$. Such a repeated equilibrium is unstable of the second order, which is called an *upper-saddle of the second order*.

(iii$_2$) For $a_0 < 0$, $\Delta \dot{x}_i < 0$ exists. So a flow of x reaches to $x^* = a_1^{(i)} \in S_1$ with $g(x^*) = a_1$ from the initial point of $x_0 > a_1^{(i)}$ and it goes to the negative infinity from $x_0 < a_1^{(i)}$. Such a repeated equilibrium is unstable of the second order, which is called a *lower-saddle of the second order*.

(iv) Under a condition of

$$\Delta = B^2 - 4AC > 0, \text{ and } dg/dx|_{x^*=a_j^{(s)}} = 0, \ d^2g/dx^2|_{x^*=a_j^{(i)}} \neq 0$$

we have

$$\dot{x} = f(x, \mathbf{p}) = a_0(g(x) - a_1)(g(x) - a_2)$$

and

$$a_1 \neq a_2,$$

$$\frac{df}{dx}\Big|_{x^*=a_i^{(s)}} = a_0(a_i - a_1)\frac{dg}{dx}\Big|_{x^*=a_i^{(s)}} + a_0(a_i - a_2)\frac{dg}{dx}\Big|_{x^*=a_i^{(s)}} = 0$$

$$\frac{d^2f}{dx^2}\Big|_{x^*=a_i^{(s)}} = a_0(a_i - a_1)\frac{d^2g}{dx^2}\Big|_{x^*=a_i^{(s)}} + 2a_0(\frac{dg}{dx}\Big|_{x^*=a_i^{(s)}})^2$$

$$+ a_0(a_i - a_2)\frac{d^2g}{dx^2}\Big|_{x^*=a_i^{(s)}}$$

$$= a_0(a_i - a_j)\frac{d^2g}{dx^2}\Big|_{x^*=a_i^{(s)}}$$

$$i, j \in \{1, 2\}, \ j \neq i.$$

Thus,

$$\Delta \dot{x}_i = \frac{1}{2!}d^2f/dx^2|_{x^*=a_j^{(i)}}\Delta x_i^2 = \frac{1}{2}a_0(a_j - a_s)d^2g/dx^2|_{x^*=a_j^{(i)}}\Delta x_i^2$$

$$s, j \in \{1, 2\}, \ s \neq j.$$

(iv$_1$) For $d^2f/dx^2|_{x^*=a_j^{(i)}} > 0$, $\Delta \dot{x}_i > 0$ exists. So a flow of x reaches to $x^* = a_\alpha^{(j)} \in S_\alpha$ repeatedly with $g(x^*) = a_\alpha$ ($\alpha = 1, 2$) from the initial point of $x_0 < a_\alpha^{(i)}$ and it goes to the positive infinity from $x_0 > a_\alpha^{(i)}$. Such a repeated equilibrium is unstable of the second order, which is called an *upper-saddle of the second order*.

(iv$_2$) For $d^2 f/dx^2|_{x^*=a_j^{(i)}} > 0$, $\Delta \dot{x}_i < 0$ exists. So a flow of x reaches to $x^* = a_\alpha^{(j)} \in S_\alpha$ repeatedly with $g(x^*) = a_\alpha$ ($\alpha = 1, 2$) from the initial point of $x_0 > a_\alpha^{(i)}$ and it goes to the negative infinity from $x_0 < a_\alpha^{(i)}$. Such a repeated equilibrium is unstable of the second order, which is called a *lower-saddle of the second order*.

(v) Under a condition of

$$\Delta = B^2 - 4AC = 0, \text{ and } dg/dx|_{x^*=a_j^{(i)}} = 0, \ d^2g/dx^2|_{x^*=a_j^{(i)}} \neq 0$$

we have

$$\dot{x} = f(x, \mathbf{p}) = a_0(g(x) - a_1)^2$$

and

$$a_1 = a_2 = a_i,$$

$$\frac{df}{dx}\Big|_{x^*=a_i^{(s)}} = a_0(a_i - a_1)\frac{dg}{dx}\Big|_{x^*=a_i^{(s)}} + a_0(a_i - a_2)\frac{dg}{dx}\Big|_{x^*=a_i^{(s)}} = 0,$$

$$\frac{d^2 f}{dx^2}\Big|_{x^*=a_i^{(s)}} = a_0(a_i - a_1)\frac{d^2g}{dx^2}\Big|_{x^*=a_i^{(s)}} + 2a_0(\frac{dg}{dx}\Big|_{x^*=a_i^{(s)}})^2$$

$$+ a_0(a_i - a_2)\frac{d^2g}{dx^2}\Big|_{x^*=a_i^{(s)}}$$

$$= 2a_0(\frac{dg}{dx}\Big|_{x^*=a_i^{(s)}})^2 = 0,$$

$$\frac{d^3 f}{dx^3}\Big|_{x^*=a_i^{(s)}} = a_0(a_i - a_1)\frac{d^3g}{dx^3}\Big|_{x^*=a_i^{(s)}} + 6a_0(\frac{dg}{dx}\Big|_{x^*=a_i^{(s)}})(\frac{d^2g}{dx^2}\Big|_{x^*=a_i^{(s)}})$$

$$+ a_0(a_i - a_2)(\frac{d^3g}{dx^3}\Big|_{x^*=a_i^{(s)}}) = 0,$$

$$\frac{d^4 f}{dx^4}\Big|_{x^*=a_i^{(s)}} = a_0(a_i - a_1)\frac{d^4g}{dx^4}\Big|_{x^*=a_i^{(s)}} + 6a_0(\frac{d^2g}{dx^2}\Big|_{x^*=a_i^{(s)}})^2$$

$$+ 8a_0(\frac{dg}{dx}\Big|_{x^*=a_i^{(s)}})(\frac{d^3g}{dx^3}\Big|_{x^*=a_i^{(s)}}) + a_0(a_i - a_2)\frac{d^4g}{dx^4}\Big|_{x^*=a_i^{(s)}}$$

$$= 6a_0(\frac{d^2g}{dx^2}\Big|_{x^*=a_i^{(s)}})^2$$

$$i, j \in \{1, 2\}, \ i \neq j.$$

Thus,

$$\Delta \dot{x}_i = \frac{1}{4!}d^4 f/dx^4\Big|_{x^*=a_j^{(i)}}\Delta x_i^4 = \frac{1}{4}a_0(d^2g/dx^2|_{x^*=a_j^{(i)}})^2\Delta x_i^4.$$

(v_1) For $a_0 > 0$ and $d^2g/dx^2|_{x^*=a_1^{(i)}} \neq 0$, $\Delta \dot{x}_i > 0$ exists. So a flow of x reaches to $x^* = a_1^{(j)} \in S_1$ repeatedly with $g(x^*) = a_1$ from the initial point of $x_0 < a_1^{(i)}$ and it goes to the positive infinity from $x_0 > a_1^{(i)}$. Such a fourth repeated equilibrium is unstable of the fourth order, which is called a fourth-order *upper-saddle* bifurcation for an onset of two second-order *upper-saddle* equilibriums.

(v_2) For $a_0 < 0$ and $d^2g/dx^2|_{x^*=a_1^{(i)}} \neq 0$, $\Delta \dot{x}_i < 0$ exists. So a flow of x reaches to the fourth repeated equilibriums of $x^* = a_1^{(j)} \in S_1$ repeatedly with $g(x^*) = a_1$ from the initial point of $x_0 > a_1^{(i)}$ and it goes to the negative infinity from $x_0 < a_1^{(i)}$. Such a fourth repeated equilibrium is unstable of the fourth order, which is called a fourth-order *lower-saddle* bifurcation for an onset of two second-order *lower-saddle* equilibriums.

The theorem is proved.   □

The stability and bifurcation of equilibriums for the 1-dimensional quadratic nonlinear functional system in Eq. (2.1) are illustrated in Fig. 2.1. The stable and unstable functional equilibriums varying with the vector parameter are depicted by the solid and dashed curves, respectively. The bifurcation point of functional equilibriums occurs at the double-repeated equilibrium at the boundary of $\mathbf{p}_0 \in \partial\Omega_{12}$. In Fig. 2.1a, for $a_0 > 0$, the relation of $g(x)$ and $\dot{x}$ with $\Delta < 0$, $\Delta = 0$, and $\Delta > 0$ are presented to show the functional equilibrium of $g(x^*) = a_i$ ($i = 1, 2$) with $\dot{x} = 0$. In Fig. 2.1b, for $a_0 > 0$, the equilibriums x^* simply relative to $g(x^*) = a_2$ and $g(x^*) = a_1$ for $\Delta > 0$ and $\frac{dg}{dx}|_{x^*} > 0$ are unstable and stable, respectively. However, in Fig. 2.1c, for $a_0 > 0$, the equilibriums x^* simply pertaining to $g(x^*) = a_2$ and $g(x^*) = a_1$ for $\Delta > 0$ and $\frac{dg}{dx}|_{x^*} < 0$ are stable and unstable, respectively. The bifurcation of equilibriums also occurs at $\Delta = 0$. The flow of $x(t)$ is a forward *upper* flow for $a_0 > 0$, and the equilibrium point x^* with $g(x^*) = -\frac{B(\mathbf{p}_0)}{2A(\mathbf{p}_0)}$ at $\Delta = 0$ is termed an *upper-saddle*. Such a bifurcation is termed an *upper-saddle-node* bifurcation. For $\Delta < 0$, no equilibrium exists. Such a 1-dimensional quadratic functional nonlinear system is termed a non-equilibrium system. For $\Delta < 0$ and $a_0 > 0$, the flow of $x(t)$ is always toward the positive direction because of $\dot{x} = a_0[(g(x) + \frac{B}{2A})^2 + (-\frac{\Delta}{4A^2})] > 0$. Similarly, in Fig. 2.1d, for $a_0 < 0$, the relation of $g(x)$ and $\dot{x}$ for $\Delta < 0$, $\Delta = 0$, and $\Delta > 0$ are presented to show the functional equilibrium of $g(x^*) = a_i$ ($i = 1, 2$) with $\dot{x} = 0$. In Fig. 2.1e, for $a_0 < 0$, the equilibriums x^* simply relative to $g(x^*) = a_2$ and $g(x^*) = a_1$ for $\Delta > 0$ and $\frac{dg}{dx}|_{x^*} > 0$ are stable and unstable, respectively. However, in Fig. 2.1f, for $a_0 < 0$, the simple equilibriums x^* for $g(x^*) = a_2$ and $g(x^*) = a_1$ with $\Delta > 0$ and $\frac{dg}{dx}|_{x^*} < 0$ are unstable and stable, respectively. The functional bifurcation of equilibriums also occurs at $\Delta = 0$. The flow of $x(t)$ is a forward *lower* flow for $a_0 < 0$, and the equilibrium point x^* with $g(x^*) = -\frac{B(\mathbf{p}_0)}{2A(\mathbf{p}_0)}$ at $\Delta = 0$ is termed an *lower-saddle*. Such a bifurcation is termed a functional *upper-saddle-node* bifurcation. For $\Delta < 0$, no equilibrium exists. Such a 1-dimensional system is termed the non-

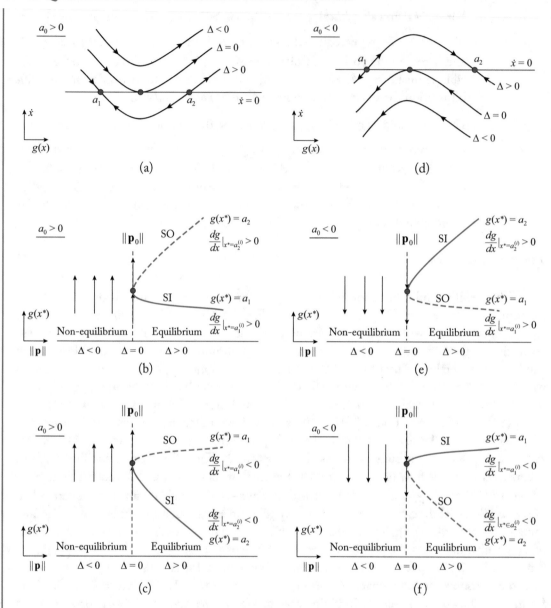

Figure 2.1: Stability and bifurcation of two functional equilibriums in the quadratic nonlinear functional dynamical system. For $a_0 > 0$, (a) phase portrait, (b, c) two *upper-saddle-node* functional bifurcations. For $a_0 < 0$, (d) phase portrait, (e, f) two *lower-saddle-node functional* bifurcations. Stable and unstable equilibriums are represented by solid and dashed curves, respectively. SO: source, SI: sink.

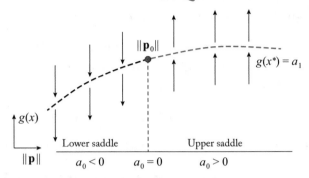

Figure 2.2: Stability and bifurcation of a double-repeated functional equilibrium of the second order in the quadratic functional dynamical system. Unstable equilibrium is represented by a dashed curve. The stability switching from the *lower-saddle* to *upper-saddle* is labeled by a circular symbol.

equilibrium system. For $\Delta < 0$ and $a_0 > 0$, the flow of $x(t)$ is always toward the negative direction because of $\dot{x} = a_0[(g(x) + \frac{B}{2A})^2 + (-\frac{\Delta}{4A^2})] < 0$.

To illustrate the stability and bifurcation of equilibrium with singularity in a 1-dimensional, quadratic functional nonlinear system, the equilibrium of $\dot{x} = a_0(g(x) - a_1)^2$ is presented in Fig. 2.2. The *upper-saddle* and *lower-saddle* of the simple equilibrium $x^* \in S_1$ of $g(x^*) = a_1$ with the second-order multiplicity are unstable, which are depicted by dashed curves. At $a_0 = 0$, the *upper-saddle* and *lower-saddle* equilibriums will be switched, which is marked by a circular symbol.

2.1.2 WITHOUT FUNCTIONAL SINGULARITY

For $\Delta > 0$, the quadratic nonlinear functional dynamical system becomes

$$\dot{x} = f(x, \mathbf{p}) = a_0(g(x) - a_1)(g(x) - a_2). \qquad (2.25)$$

(i) For a linear function of $g(x) = \alpha_0 x + \alpha_1$, we have

$$\dot{x} = f(x, \mathbf{p}) = a_0 \alpha_0^2 (x - b_1)(x - b_2) \qquad (2.26)$$

if

$$\Delta > 0, \ \{b_1, b_2\} = \text{sort}\{a_1^{(1)}, a_2^{(1)}\}, \ b_1 > b_2;$$
$$a_i^{(1)} = \frac{1}{\alpha_0}(a_i - \alpha_1), \ i = 1, 2. \qquad (2.27)$$

(ii) For a quadratic nonlinear function of $g(x) = \alpha_0 x^2 + \alpha_1 x + \alpha_2$, we have

$$g(x) = a_i \implies \alpha_0 x^2 + \alpha_1 x + \alpha_2 = a_i$$
$$\Delta_i = \alpha_1^2 - 4\alpha_0(\alpha_2 - a_i). \tag{2.28}$$

(ii$_1$) If

$$\Delta > 0, \ \Delta_i < 0, \ i = 1, 2, \tag{2.29}$$

then the quadratic nonlinear functional dynamical system becomes

$$\dot{x} = f(x, \mathbf{p}) = a_0(g(x) - a_1)(g(x) - a_2) \tag{2.30}$$

which is without any equilibriums. In other words, $g(x^*) = a_i$ ($i = 1, 2$) do not have any real solutions of x^* to make $\dot{x} = 0$.

(ii$_{1a}$) If

$$a_0(g(x) - a_1)(g(x) - a_2) > 0, \tag{2.31}$$

then the flow of the quadratic nonlinear functional dynamical system is called a positive flow.

(ii$_{1b}$) If

$$a_0(g(x) - a_1)(g(x) - a_2) < 0, \tag{2.32}$$

then the flow of the quadratic nonlinear functional dynamical system is called a negative flow.

(ii$_2$) If

$$\Delta > 0, \ \Delta_i > 0, \ \Delta_j < 0, \ i, j \in \{1, 2\}, \ j \neq i, \tag{2.33}$$

then the quadratic nonlinear functional dynamical system becomes

$$\dot{x} = f(x, \mathbf{p}) = a_0\alpha_0(g(x) - a_j)(x - b_1)(x - b_2), \tag{2.34}$$

where

$$\{b_1, b_2\} = \text{sort}\{a_i^{(1)}, a_i^{(2)}\}, \ b_1 > b_2;$$
$$a_i^{(1,2)} = -\frac{1}{2\alpha_0}(\alpha_1 \mp \sqrt{\Delta_i}), \ i, j \in \{1, 2\}; \ j \neq i. \tag{2.35}$$

(ii$_3$) If

$$\Delta > 0 \text{ and } \Delta_i > 0, \ i = 1, 2, \tag{2.36}$$

then the quadratic nonlinear functional dynamical system becomes

$$\dot{x} = f(x, \mathbf{p}) = a_0\alpha_0^2(x - b_1)(x - b_2)(x - b_3)(x - b_4), \tag{2.37}$$

where

$$\{b_1, b_2, b_3, b_4\} = \text{sort}\{\cup_{i=1}^2 \{a_i^{(1)}, a_i^{(2)}\}\}, \ b_{s+1} > b_s;$$
$$a_i^{(1,2)} = -\frac{1}{2\alpha_0}(\alpha_1 \mp \sqrt{\Delta_i}), \ i = 1, 2. \tag{2.38}$$

(ii$_4$) If

$$\Delta > 0 \text{ and } \Delta_i > 0, \ \Delta_j = 0, \ \text{for } i, j \in \{1, 2\}, \ j \neq i, \tag{2.39}$$

then the quadratic nonlinear functional dynamical system becomes

$$\dot{x} = f(x, \mathbf{p}) = a_0 \alpha_0^2 (x - b_{i_1})(x - b_{i_2})(x - b_{i_3})^2, \tag{2.40}$$

where

$$\{b_1, b_2, b_3\} = \text{sort}\{\{a_i^{(1)}, a_i^{(2)}\}, a_j^{(1)} = a_j^{(2)}\} \ b_{s+1} \geq b_s;$$
$$a_i^{(1,2)} = -\frac{1}{2\alpha_0}(\alpha_1 \mp \sqrt{\Delta_i}), \ a_j^{(1,2)} = -\frac{1}{2\alpha_0}\alpha_1, \ i, j \in \{1, 2\}, \tag{2.41}$$
$$j \neq i; \ i_k \in \{1, 2, 3\}, \ k = 1, 2, 3.$$

(iii) For a cubic nonlinear function of $g(x) = \alpha_0 x^3 + \alpha_1 x^2 + \alpha_2 x + \alpha_3$, we have

$$g(x) = a_i \ \Rightarrow \ \alpha_0 x^3 + \alpha_1 x^2 + \alpha_2 x + \alpha_3 = a_i$$
$$\alpha_0 x^3 + \alpha_1 x^2 + \alpha_2 x + \alpha_3 - a_i = \alpha_0(x - a_i^{(1)})(x^2 + \beta_i^{(1)} x + \beta_i^{(2)}) \tag{2.42}$$
$$\Delta_i = (\beta_i^{(1)})^2 - 4\beta_i^{(2)}.$$

(iii$_1$) If

$$\Delta > 0, \ \Delta_i < 0, \ i = 1, 2, \tag{2.43}$$

then the quadratic nonlinear functional dynamical system has two simple equilibriums, i.e.,

$$\dot{x} = f(x, \mathbf{p}) = a_0 \alpha_0^2 \prod_{i=1}^2 [(x + \frac{1}{2}\beta_i^{(1)})^2 + \frac{1}{4}(-\Delta_i)](x - b_1)(x - b_2), \tag{2.44}$$

where

$$\{b_1, b_2\} = \text{sort}\{a_1^{(1)}, a_2^{(1)}\}. \tag{2.45}$$

(iii$_2$) If

$$\Delta > 0, \ \Delta_i > 0, \ \Delta_j < 0, \ i, j \in \{1, 2\}, \ j \neq i, \tag{2.46}$$

then the quadratic nonlinear functional dynamical system has four simple equilibriums, i.e.,

$$\dot{x} = f(x, \mathbf{p}) = a_0 \alpha_0^2 [(x - \frac{1}{2}\beta_i^{(1)})^2 + \frac{1}{4}(-\Delta_i)] \prod_{s=1}^{4}(x - b_s), \qquad (2.47)$$

where

$$\{\cup_{s=1}^{4} b_s\} = \text{sort}\{\cup_{k=1}^{3} a_i^{(k)}, a_j^{(1)}\}, \ b_{s+1} > b_s;$$
$$a_i^{(2,3)} = -(\frac{1}{2}\beta_i^{(1)} \mp \sqrt{\Delta_i}), \ i, j \in \{1, 2\}; \ j \neq i. \qquad (2.48)$$

(iii$_3$) If

$$\Delta > 0 \text{ and } \Delta_i > 0, \ i = 1, 2, \qquad (2.49)$$

then the quadratic nonlinear functional dynamical system has six simple equilibriums, i.e.,

$$\dot{x} = f(x, \mathbf{p}) = a_0 \alpha_0^2 \prod_{s=1}^{6}(x - b_s), \qquad (2.50)$$

where

$$\{\cup_{s=1}^{6} b_s\} = \text{sort}\{\cup_{k=1}^{3} a_1^{(k)}, \cup_{k=1}^{3} a_2^{(k)}\}, \ b_{s+1} > b_s;$$
$$a_i^{(2,3)} = -(\frac{1}{2}\beta_i^{(1)} \mp \sqrt{\Delta_i}), \ i = 1, 2. \qquad (2.51)$$

(iii$_4$) If

$$\Delta > 0 \text{ and } \Delta_i > 0, \ \Delta_j = 0, \ \text{for } i, j \in \{1, 2\}, j \neq i, \qquad (2.52)$$

then the quadratic nonlinear functional dynamical system has four simple equilibriums plus one repeated equilibrium, i.e.,

$$\dot{x} = f(x, \mathbf{p}) = a_0 \alpha_0^2 \prod_{s=1}^{4}(x - b_{i_s})(x - b_{i_5})^2, \qquad (2.53)$$

where

$$\{\cup_{s=1}^{5} b_s\} = \text{sort}\{\cup_{k=1}^{3} a_i^{(k)}, a_j^{(1)}, a_j^{(2)} = a_j^{(3)}\}, \ b_{s+1} > b_s;$$
$$a_i^{(2,3)} = -(\frac{1}{2}\beta_i^{(1)} \mp \sqrt{\Delta_i}), \ a_j^{(2,3)} = -\frac{1}{2}\beta_j^{(1)}, \ i, j \in \{1, 2\}, \ j \neq i; \quad (2.54)$$
$$i_k \in \{1, 2, \ldots, 5\}, \ k = 1, 2, \ldots, 5.$$

(iii$_5$) If

$$\Delta > 0 \text{ and } \Delta_i > 0, \ \Delta_j = 0, \text{ with } a_j^{(1)} = a_j^{(2)} = a_j^{(3)}$$
$$\text{for } i, j \in \{1, 2\}, \ j \neq i, \tag{2.55}$$

then the quadratic nonlinear functional dynamical system has three simple equilibriums plus one triple-repeated equilibrium, i.e.,

$$\dot{x} = f(x, \mathbf{p}) = a_0 \alpha_0^2 \prod_{s=1}^{3} (x - b_{i_s})(x - b_{i_5})^3, \tag{2.56}$$

where

$$\{\cup_{s=1}^4 b_s\} = \text{sort} \{\cup_{k=1}^3 a_i^{(k)}, a_j^{(1)}, a_j^{(2)} = a_j^{(3)}\}, \ b_{s+1} > b_s;$$
$$a_i^{(2,3)} = -(\frac{1}{2}\beta_i^{(1)} \mp \sqrt{\Delta_i}), \ a_j^{(2,3)} = -\frac{1}{2}\beta_j^{(1)}, \ i, j \in \{1, 2\}, \ j \neq i; \tag{2.57}$$
$$i_k \in \{1, 2, \ldots, 4\}, \ k = 1, 2, \ldots, 4.$$

(iii$_6$) If
$$\Delta > 0 \text{ and } \Delta_i = 0, \ \Delta_j = 0, \text{ for } i, j \in \{1, 2\}, \ j \neq i, \tag{2.58}$$

then the quadratic nonlinear functional dynamical system has two simple equilibriums and two repeated equilibriums, i.e.,

$$\dot{x} = f(x, \mathbf{p}) = a_0 \alpha_0^2 (x - b_{i_1})(x - b_{i_2})^2 (x - b_{i_3})(x - b_{i_4})^2, \tag{2.59}$$

where

$$\{\cup_{s=1}^4 b_s\} = \text{sort} \{\cup_{i=1}^2 \{a_i^{(1)}, a_i^{(2)} = a_i^{(3)}\}\}, \ b_{s+1} > b_s;$$
$$a_l^{(s)} = -\frac{1}{2}\beta_l^{(1)}, \ l = 1, 2; \ i, j \in \{1, 2\}, \ j \neq i; \ i_k \in \{1, 2, \ldots, 4\}, \tag{2.60}$$
$$k = 1, 2, \ldots, 4.$$

(iii$_7$) If

$$\Delta > 0 \text{ and } \Delta_i = 0, \ \Delta_j = 0, \text{ with } a_j^{(1)} = a_j^{(2)} = a_j^{(3)}$$
$$\text{for } i, j \in \{1, 2\}, \ j \neq i, \tag{2.61}$$

then the quadratic nonlinear functional dynamical system has one simple equilibrium, one double-repeated equilibrium, and one triple-repeated equilibrium, i.e.,

$$\dot{x} = f(x, \mathbf{p}) = a_0 \alpha_0^2 (x - b_{i_1})(x - b_{i_2})^2 (x - b_{i_3})^3, \tag{2.62}$$

where

$$\{\cup_{s=1}^{3} b_s\} = \text{sort}\{a_i^{(1)}, a_i^{(2)} = a_i^{(3)}, a_j^{(1)} = a_j^{(2)} = a_j^{(3)}\}, \ b_{s+1} > b_s;$$

$$a_l^{(s)} = -\frac{1}{2}\beta_l^{(1)}, \ l = 1, 2; \ i, j \in \{1, 2\}, \ j \neq i; \ i_k \in \{1, 2, 3\}, \ k = 1, 2, 3.$$

(2.63)

(iii$_8$) If

$$\Delta > 0 \text{ and } \Delta_i = 0, \text{ with } a_i^{(1)} = a_i^{(2)} = a_i^{(3)} \text{ for } i = 1, 2, \qquad (2.64)$$

then the quadratic nonlinear functional dynamical system has two triple-repeated equilibriums, i.e.,

$$\dot{x} = f(x, \mathbf{p}) = a_0 \alpha_0^2 (x - b_{i_1})^3 (x - b_{i_2})^3, \qquad (2.65)$$

where

$$\{\cup_{s=1}^{2} b_s\} = \text{sort}\{\cup_{i=1}^{3}\{a_i^{(1)} = a_i^{(2)} = a_i^{(3)}\}\}, \ b_{s+1} > b_s;$$

$$a_i^{(s)} = -\frac{1}{2}\beta_i^{(1)}, \ i = 1, 2; \ i_k \in \{1, 2, \}, \ k = 1, 2.$$

(2.66)

For other polynomial functions, the similar discussion can be done. The illustrations of multiple-equilibriums system with higher-order singularity can be found from Luo (2020).[1]

2.1.3   WITH FUNCTIONAL SINGULARITY

For $\Delta = 0$, the quadratic nonlinear functional dynamical system becomes

$$\dot{x} = f(x, \mathbf{p}) = a_0 (g(x) - a_1)^2. \qquad (2.67)$$

(i) For a linear function of $g(x) = \alpha_0 x + \alpha_1$, we have

$$\dot{x} = f(x, \mathbf{p}) = a_0 \alpha_0^2 (x - b_1)^2, \qquad (2.68)$$

where

$$\Delta = 0, \ b_1 = \frac{1}{\alpha_0}(a_1 - \alpha_1). \qquad (2.69)$$

- For $a_0 > 0$, the quadratic nonlinear functional dynamical system has an *upper-saddle* of the second order.

- For $a_0 < 0$, the quadratic nonlinear functional dynamical system has a *lower-saddle* of the second order.

[1]Albert C. J. Luo (2020), *Bifurcation and Stability in Nonlinear Dynamical Systems*, Springer, Switzerland.

(ii) For a quadratic nonlinear function of $g(x) = \alpha_0 x^2 + \alpha_1 x + \alpha_2$, we have

$$g(x) = a_1 \Rightarrow \alpha_0 x^2 + \alpha_1 x + \alpha_2 = a_1$$
$$\Delta_1 = \alpha_1^2 - 4\alpha_0(\alpha_2 - a_1). \tag{2.70}$$

(ii$_1$) If

$$\Delta = 0, \ \Delta_1 < 0, \tag{2.71}$$

then the quadratic nonlinear functional dynamical system is

$$\dot{x} = f(x, \mathbf{p}) = a_0(g(x) - a_1)^2, \tag{2.72}$$

which is without any equilibriums.

- For $a_0 > 0$, the flow of the quadratic nonlinear functional dynamical system is called a positive flow.
- For $a_0 < 0$, the flow of the quadratic nonlinear functional dynamical system is called a negative flow.

(ii$_2$) If

$$\Delta = 0, \ \Delta_1 > 0, \tag{2.73}$$

then the quadratic nonlinear functional dynamical system is

$$\dot{x} = f(x, \mathbf{p}) = a_0 \alpha_0^2 (x - b_1)^2 (x - b_2)^2, \tag{2.74}$$

where

$$\{b_1, b_2\} = \text{sort}\{a_1^{(1)}, a_1^{(2)}\}, \ b_1 < b_2; \ a_1^{(1,2)} = \frac{1}{2\alpha_0}(\alpha_1 \pm \sqrt{\Delta_1}). \tag{2.75}$$

- For $a_0 > 0$, the quadratic nonlinear functional dynamical system has two *upper-saddle*.
- For $a_0 < 0$, the quadratic nonlinear functional dynamical system has two *lower-saddles*.

(ii$_3$) If

$$\Delta = 0 \text{ and } \Delta_1 = 0, \tag{2.76}$$

then the quadratic nonlinear functional dynamical system is

$$\dot{x} = f(x, \mathbf{p}) = a_0 \alpha_0^2 (x - b_1)^4, \tag{2.77}$$

where

$$b_1 = a_i^{(1)} = a_i^{(2)} = -\frac{1}{2\alpha_0}\alpha_1. \tag{2.78}$$

- For $a_0 > 0$, the quadratic nonlinear functional dynamical system has a fourth-order *upper-saddle*.
- For $a_0 < 0$, the quadratic nonlinear functional dynamical system has a fourth-order *lower-saddle*.

(iii) For a cubic nonlinear function of $g(x) = \alpha_0 x^3 + \alpha_1 x^2 + \alpha_2 x + \alpha_3$, we have

$$g(x) = a_1 \;\Rightarrow\; \alpha_0 x^3 + \alpha_1 x^2 + \alpha_2 x + \alpha_3 = a_1$$
$$\alpha_0 x^3 + \alpha_1 x^2 + \alpha_2 x + \alpha_3 - a_1 = \alpha_0(x - a_1^{(1)})(x^2 + \beta_1^{(1)} x + \beta_1^{(2)}) \qquad (2.79)$$
$$\Delta_1 = (\beta_1^{(1)})^2 - 4\beta_1^{(2)}.$$

(iii$_1$) If

$$\Delta = 0, \;\; \Delta_i < 0, \;\; i = 1, 2, \qquad (2.80)$$

then the quadratic nonlinear functional dynamical system is

$$\dot{x} = f(x, \mathbf{p}) = a_0 \alpha_0^2 [(x + \tfrac{1}{2}\beta_1^{(1)})^2 + \tfrac{1}{4}(-\Delta_1)]^2 (x - b_1)^2, \qquad (2.81)$$

where

$$b_1 = a_1^{(1)}. \qquad (2.82)$$

- For $a_0 > 0$, the functional dynamical system has an *upper-saddle*.
- For $a_0 < 0$, the functional dynamical system has a *lower-saddle*.

(iii$_2$) If

$$\Delta = 0, \;\; \Delta_1 > 0, \qquad (2.83)$$

then the quadratic nonlinear functional dynamical system is

$$\dot{x} = f(x, \mathbf{p}) = a_0 \alpha_0^2 \prod_{i=1}^{3} (x - b_i)^2, \qquad (2.84)$$

where

$$\{\cup_{s=1}^3 b_s\} = \text{sort}\,\{\cup_{k=2}^3 a_i^{(k)}, a_j^{(1)}\}, \;\; b_{s+1} > b_s;$$
$$a_i^{(2,3)} = -(\tfrac{1}{2}\beta_i^{(1)} \mp \sqrt{\Delta_i}), \;\; i, j \in \{1, 2\}; \;\; j \neq i. \qquad (2.85)$$

- For $a_0 > 0$, the quadratic nonlinear functional dynamical system has three *upper-saddles*.
- For $a_0 < 0$, the quadratic nonlinear functional dynamical system has three *lower-saddles*.

(iii$_3$) If

$$\Delta = 0, \text{ and } \Delta_1 = 0; \tag{2.86}$$

then the quadratic nonlinear functional dynamical system is

$$\dot{x} = f(x, \mathbf{p}) = a_0 \alpha_0^2 (x - b_{i_1})^2 (x - b_{i_1})^4, \tag{2.87}$$

where

$$\{\cup_{s=1}^2 b_s\} = \text{sort}\{a_1^{(1)}, a_1^{(2)} = a_1^{(3)}\}, \ b_{s+1} > b_s; \tag{2.88}$$
$$a_1^{(2,3)} = -\frac{1}{2}\beta_1^{(1)}, \ i_k \in \{1, 2, \}, \ k = 1, 2.$$

- For $a_0 > 0$, the functional dynamical system has one second-order *upper-saddle*, and one fourth-order *upper-saddle*.
- For $a_0 < 0$, the functional dynamical system has one second-order *lower-saddle*, and one fourth-order *lower-saddle*.

(iii$_4$) If

$$\Delta = 0 \text{ and } \Delta_1 = 0 \text{ with } a_1^{(1)} = a_1^{(2)} = a_1^{(3)}, \tag{2.89}$$

then the quadratic nonlinear functional dynamical system is

$$\dot{x} = f(x, \mathbf{p}) = a_0 \alpha_0^2 (x - b_1)^6, \tag{2.90}$$

where

$$b_1 = a_1^{(1)} = a_1^{(2,3)} = -\frac{1}{2}\beta_1^{(1)}. \tag{2.91}$$

- For $a_0 > 0$, the functional dynamical system has one sixth-order *upper-saddle*.
- For $a_0 < 0$, the functional dynamical system has one sixth-order *lower-saddle*.

Consider $g(x)$ being other polynomial functions, the similar discussion can be completed as an exercise for readers. If $g(x)$ is a (2m)th-degree or (2m+1)th-degree polynomial function, the singularity of the corresponding equilibriums can be discussed in Luo (2020).[2] For $g(x)$ being a non-polynomial function, the Taylor series expansion of $g(x)$ should be developed at the specific equilibriums for existence and singularity of equilibriums.

To illustrate equilibriums in the quadratic functional dynamical system, consider $\Delta = B^2 - 4AC$ and a quadratic function of $g(x^*) = \alpha_0(x^*)^2 + \alpha_1 x^* + \alpha_2 = a_i$ with $\Delta_i = \alpha_1^2 - 4\alpha_0(\alpha_2 - a_i)$ $(i = 1, 2)$ and $a_2 \geq a_1$. For $||\mathbf{p}_{cr}||_{\Delta=0} > ||\mathbf{p}_{cr}||_{\Delta_i=0}$ $(i = 1, 2)$, four equilibriums in the quadratic functional dynamical system with quadratic function are presented in Fig. 2.3. Before the disappearance of the function $g(x) = a_i$ $(i = 1, 2)$, $f(x) = A_0[g(x)]^2 + A_1 g(x) + A_2 = 0$ does not have real solutions. The equilibriums with a functional *upper-saddle-node* bifurcation $(a_0 > 0, \Delta = 0)$ and a functional *lower-saddle-node* bifurcation $(a_0 < 0, \Delta = 0)$ are

---

[2] Albert C. J. Luo (2020), *Bifurcation and Stability in Nonline Dynamical Systems*, Springer, Switzerland.

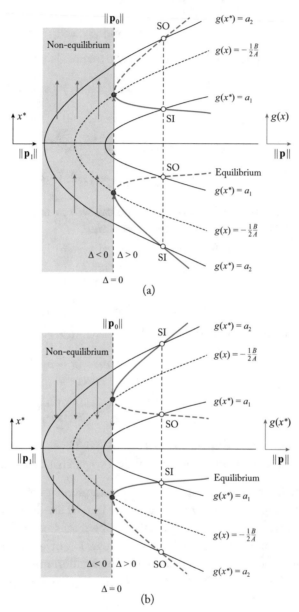

Figure 2.3: Four equilibriums in the quadratic functional dynamical system with a quadratic function: (a) a functional *upper-saddle-node* bifurcation ($a_0 > 0, \Delta = 0$), (b) a functional *lower-saddle-node* bifurcation ($a_0 < 0, \Delta = 0$). Stable and unstable equilibriums are depicted by thick solid and dashed curves, respectively. SO: source, SI: sink. The thin curves are for $g(x^*) = a_i$ ($i = 1, 2$).

sketched in Figs. 2.3a,b, respectively. Stable and unstable equilibriums are represented by thick solid and dashed curves, respectively. The acronyms SO and SI are for source and sink equilibriums. The acronyms US and LS are for *upper-* and *lower-saddle* equilibriums, respectively. The thin curves are for the quadratic functions of $g(x^*) = a_i$ $(i = 1, 2)$. For such a quadratic function in the quadratic functional system, there are four simple equilibriums and two double-repeated equilibriums.

For $||\mathbf{p}_{cr}||_{\Delta=0} < ||\mathbf{p}_{cr}||_{\Delta_i=0}$ $(i = 1, 2)$, two or four equilibriums in the quadratic functional dynamical system with a quadratic function are presented in Fig. 2.4. The equilibriums with an *upper-saddle-node* bifurcation $(a_0 > 0, \Delta_2 = 0)$ and a *lower-saddle-node* bifurcation $(a_0 > 0, \Delta_1 = 0)$ are sketched in Fig. 2.4a. The equilibriums with a *lower-saddle-node* bifurcation $(a_0 < 0, \Delta_2 = 0)$ and an *upper-saddle-node* bifurcation $(a_0 < 0, \Delta_1 = 0)$ are sketched in Fig. 2.4b. Stable and unstable equilibriums are also represented by thick solid and dashed curves, respectively. The acronyms SO and SI are for source and sink equilibriums. The acronyms US and LS are for *upper-* and *lower-saddle* equilibriums, respectively. For such a quadratic function in the quadratic functional system, there are two or four simple equilibriums and two double-repeated equilibriums.

If $f(x^*) = A_0[g(x^*)]^2 + A_1 g(x^*) + A_2 = 0$ with $\Delta \geq 0$ has two real solutions of $g(x^*) = \alpha_0 x^{*2} + \alpha_1 x^* + \alpha_2 = a_i$ $(i = 1, 2)$, the equation of $g(x^*) = a_i$ $(i \in \{1, 2\})$ does not have any solution existence due to $\Delta_i < 0$ $(i = 1, 2)$. For this case, the flow of dynamical system will be positive or negative flow. For $\Delta \geq 0$, if $\Delta_i \geq 0$ and $\Delta_j < 0$ $(i, j \in \{1, 2\}, j \neq i)$, the dynamical system has one to two equilibriums. For $\Delta \geq 0$, if $\Delta_i \geq 0$ $(i = 1, 2)$, the dynamical system has one to four equilibriums.

2.2   SWITCHING BIFURCATIONS

Definition 2.3   Consider a 1-dimensional quadratic nonlinear functional dynamical system in Eq. (2.1) as

$$\dot{x} = A(\mathbf{p})(g(x))^2 + B(\mathbf{p})g(x) + C(\mathbf{p})$$
$$= a_0(\mathbf{p})(g(x) - a(\mathbf{p}))(g(x) - b(\mathbf{p})). \tag{2.92}$$

(i) For $a < b$, the corresponding standard functional form is

$$\dot{x} = a_0(g(x) - a)(g(x) - b) \tag{2.93}$$

with two equilibrium sets of

$$x^* \in S_1 = \{a_1^{(i)} | g(a_1^{(i)}) = a_1 = a, \ i = 1, 2, \dots, N_1\} \cup \{\emptyset\}$$
$$x^* \in S_2 = \{a_2^{(i)} | g(a_2^{(i)}) = a_2 = b, \ i = 1, 2, \dots, N_2\} \cup \{\emptyset\} \tag{2.94}$$
$$\text{with } \Delta = a_0^2(a - b)^2 > 0.$$

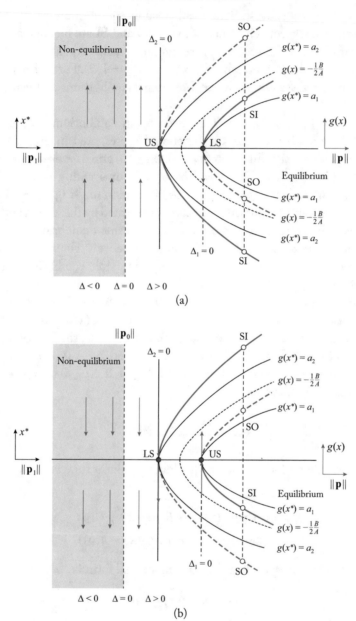

Figure 2.4: Equilibriums in the quadratic functional dynamical system with a quadratic function: (a) an *upper-saddle-node* bifurcation ($a_0 > 0, \Delta_2 = 0$), (b) a *lower-saddle-node* bifurcation ($a_0 < 0, \Delta_2 = 0$). Stable and unstable equilibriums are represented by solid and dashed curves, respectively. SO: source, SI: sink. The thin curves are for $g(x^*) = a_i$ ($i = 1, 2$).

(ii) For $a > b$, the corresponding standard functional form is

$$\dot{x} = a_0[g(x) - b][g(x) - a] \tag{2.95}$$

with two equilibrium sets of

$$x^* \in S_1 = \{a_1^{(i)} | g(a_1^{(i)}) = a_1 = b, \ i = 1, 2, \ldots, N_1\} \cup \{\emptyset\}$$
$$x^* \in S_2 = \{a_2^{(i)} | g(a_2^{(i)}) = a_2 = a, \ i = 1, 2, \ldots, N_2\} \cup \{\emptyset\} \tag{2.96}$$
$$\text{with } \Delta = a_0^2(a - b)^2 > 0.$$

(iii) For $a = b$, the corresponding standard functional form is

$$\dot{x} = a_0(g(x) - a)^2 \tag{2.97}$$

with a double-repeated functional equilibrium set of

$$x^* \in S_1 = \{a_1^{(i)} | g(a_1^{(i)}) = a_1 = a, \ i = 1, 2, \ldots, N_1\} \cup \{\emptyset\}$$
$$\text{with } \Delta = a_0^2(a - b)^2 = 0. \tag{2.98}$$

Such an equilibrium point is called a functional *saddle* of the second order.

(iii$_1$) If $a_0 > 0$, the equilibrium is a functional *upper-saddle* of the second order.

(iii$_2$) If $a_0 < 0$, the equilibrium is a functional *lower-saddle* of the second order.

(iv) The double-repeated functional equilibrium of $x^* \in S_1$ with $g(x^*) = a$ for two equilibriums switching is called a functional *saddle-node* bifurcation point of equilibrium at a point $\mathbf{p} = \mathbf{p}_0 \in \partial\Omega_{12}$, and the functional bifurcation condition is

$$\Delta = a_0^2(a - b)^2 = 0 \text{ or } a = b. \tag{2.99}$$

Theorem 2.4

(i) Under a condition of

$$a < b \text{ and } \Delta = a_0^2(a - b)^2 > 0 \tag{2.100}$$

a standard form of the quadratic nonlinear functional dynamical system in Eq. (2.93) *is*

$$\dot{x} = f(x, \mathbf{p}) = a_0(g(x) - a)(g(x) - b). \tag{2.101}$$

(i$_1$) The simple equilibrium of $x^* \in S_1$ with $g(x^*) = a$ is stable with $df/dx|_{x^*=a_1^{(i)}} < 0$ and the simple equilibrium of $x^* \in S_2$ with $g(x^*) = b$ is unstable with $df/dx|_{x^*=a_2^{(i)}} > 0$.

(i$_2$) *The simple equilibrium of $x^* \in S_1$ with $g(x^*) = a$ is unstable with $df/dx|_{x^*=a_1^{(1)}} > 0$ and the simple equilibrium of $x^* \in S_2$ with $g(x^*) = b$ is stable with $df/dx|_{x^*=a_2^{(i)}} < 0$.*

(ii) *Under a condition of*

$$a > b \text{ and } \Delta = a_0^2(a - b)^2 > 0 \tag{2.102}$$

a standard functional form of the quadratic functional system in Eq. (2.93) *is*

$$\dot{x} = a_0(g(x) - b)(g(x) - a). \tag{2.103}$$

(ii$_1$) *The simple equilibrium of $x^* \in S_2$ with $g(x^*) = a$ is unstable with $df/dx|_{x^*=a_2^{(i)}} > 0$ and the simple equilibrium of $x^* \in S_1$ with $g(x^*) = b$ is stable with $df/dx|_{x^*=a_1^{(i)}} < 0$.*

(ii$_2$) *The simple equilibrium $x^* \in S_2$ with $g(x^*) = a$ is stable with $df/dx|_{x^*=a_2^{(i)}} < 0$ and the simple equilibrium of $x^* \in S_1$ with $g(x^*) = b$ is unstable with $df/dx|_{x^*=a_1^{(i)}} > 0$.*

(iii) *For $a = b$, the corresponding standard functional form with $\Delta = 0$ is*

$$\dot{x} = f(x, \mathbf{p}) = a_0(g(x) - a)^2. \tag{2.104}$$

(iii$_1$) *If $a_0(\mathbf{p}) > 0$ and $dg/dx|_{x^*=a_1^{(i)}} \neq 0$, then the equilibrium of $x^* \in S_1$ with $g(x^*) = a$ is a functional upper-saddle of the second order with $d^2f/dx^2|_{x^*=a_1^{(i)}} > 0$. The equilibrium of $x^* \in S_1$ with $g(x^*) = a$ for two equilibriums switching is a functional upper-saddle-node bifurcation of the second order.*

(iii$_2$) *If $a_0(\mathbf{p}) < 0$ and $dg/dx|_{x^*=a_1^{(i)}} \neq 0$, then the equilibrium of $x^* \in S_1$ with $g(x^*) = a$ is a functional lower-saddle of the second order with $d^2f/dx^2|_{x^*=a} < 0$. The equilibrium of $x^* \in S_1$ with $g(x^*) = a$ for two equilibriums switching is a functional lower-saddle-node bifurcation of the second order.*

(iv) *Under a condition of*

$$\Delta = a_0^2(a - b)^2 > 0 \text{ and } dg/dx|_{x^* \in S_j} = 0, \ d^2g/dx^2|_{x^*=a_j^{(i)}} \neq 0 \ (j \in \{1, 2\})$$

$$\{a_1, a_2\} = sort\{a, b\}, \ a_2 > a_1; \tag{2.105}$$

a standard functional form of the quadratic functional system in Eq. (2.93) *is*

$$\dot{x} = f(x, \mathbf{p}) = a_0(g(x) - a_1)(g(x) - a_2). \tag{2.106}$$

(iv_1) If $d^2 f/dx^2|_{x^*=a_j^{(i)}} > 0$, the repeated equilibrium of $x^* = a_j^{(i)} \in S_j$ ($j \in \{1, 2\}$) with $g(x^*) = a_j$ is an upper-saddle of the second order. The bifurcation at the repeated equilibrium of $x^* = a_j^{(i)} \in S_j$ for the switching of two simple functional equilibriums of $g(x^*) = a_j$ is called an upper-saddle node bifurcation.

(iv_2) If $d^2 f/dx^2|_{x=a_j^{(i)}} < 0$, the repeated equilibrium of $x^* = a_j^{(i)} \in S_j$ ($j \in \{1, 2\}$) with $g(x^*) = a_j$ is a lower-saddle of the second order. The bifurcation at the repeated equilibrium of $x^* = a_j^{(i)} \in S_j$ ($j = 1, 2$) for the switching of two simple functional equilibriums of $g(x^*) = a_j$ is called a lower-saddle node bifurcation.

(v) Under a condition of

$$\Delta = a_0^2(a - b)^2 = 0 \text{ and } dg/dx|_{x^* \in S_1} = 0 \text{ with } a_1 = a = b, \tag{2.107}$$

a standard form of the functional dynamical system in Eq. (2.93) is

$$\dot{x} = f(x, \mathbf{p}) = a_0(g(x) - a_1)^2. \tag{2.108}$$

(v_1) If $a_0 > 0$ and $d^2 g/dx^2|_{x^*=a_1^{(i)}} \neq 0$, the repeated equilibrium of $x^* = a_1^{(i)} \in S_1$ with $g(x^*) = a_1(\mathbf{p})$ is an upper-saddle of the fourth order. The bifurcation at the repeated equilibrium of $x^* = a_1^{(i)} \in S_1$ for the switching of two second-order upper-saddle functional equilibriums of $g(x^*) = a_1$ is called a fourth-order upper-saddle-node bifurcation.

(v_2) If $a_0 < 0$ and $d^2 g/dx^2|_{x^*=a_1^{(i)}} \neq 0$, the repeated equilibrium of $x^* = a_1^{(i)} \in S_1$ with $g(x^*) = a_1(\mathbf{p})$ is an lower-saddle of the fourth order. The bifurcation at the repeated equilibrium of $x^* = a_1^{(i)} \in S_1$ for the switching of two second-order lower-saddle functional equilibriums of $g(x^*) = a_1$ is called a fourth-order lower-saddle-node bifurcation.

Proof. The theorem can be proved as for Theorem 2.2. □

Definition 2.5 If $C(\mathbf{p}) = 0$ in Eq. (2.1), a 1-dimensional quadratic functional dynamical system is

$$\dot{x} = A(\mathbf{p})(g(x))^2 + B(\mathbf{p})g(x). \tag{2.109}$$

(i) For $A(\mathbf{p}) \times B(\mathbf{p}) < 0$, the corresponding standard functional form is

$$\dot{x} = a_0 g(x)(g(x) - a) \tag{2.110}$$

with two equilibriums

$$S_1 = \{a_1^{(i)}|g(a_1^{(i)}) = a_1 = 0, \ i = 1, 2, \ldots, N_1\} \cup \{\emptyset\},$$
$$S_2 = \{a_2^{(i)}|g(a_2^{(i)}) = a_2 = a > 0, \ i = 1, 2, \ldots, N_2\} \cup \{\emptyset\}, \tag{2.111}$$

$$\text{with } a_0 = A(\mathbf{p}) \text{ and } a = -\frac{B(\mathbf{p})}{A(\mathbf{p})}.$$

(ii) For $A(\mathbf{p}) \times B(\mathbf{p}) > 0$, the corresponding standard functional form is

$$\dot{x} = a_0(g(x) - a)g(x) \tag{2.112}$$

with two equilibrium sets of

$$S_1 = \{a_1^{(i)}|g(a_1^{(i)}) = a_1 = a < 0, \ i = 1, 2, \ldots, N_2\} \cup \{\emptyset\},$$
$$S_2 = \{a_2^{(i)}|g(a_2^{(i)}) = a_2 = 0, \ i = 1, 2, \ldots, N_1\} \cup \{\emptyset\}. \tag{2.113}$$

(iii) For $B(\mathbf{p}) = 0$, the corresponding standard functional form is

$$\dot{x} = a_0(g(x))^2 \tag{2.114}$$

with a double-repeated equilibrium of $x^* = a_1^{(i)} \in S_1$ with $g(x^*) = a_1 = 0$. Such an equilibrium is called a *saddle* of the second order. If $a_0 > 0$, the equilibrium is an *upper-saddle* of the second order. If $a_0 < 0$, the equilibrium is a *lower-saddle* of the second order.

(iv) The bifurcation of $x^* = a_1^{(i)} \in S_1$ with $g(x^*) = a_1 = 0$ for two functional equilibriums switching is called a *saddle-node* bifurcation at a point $\mathbf{p} = \mathbf{p}_0 \in \partial\Omega_{12}$, and the bifurcation condition is

$$B(\mathbf{p}_0) = 0. \tag{2.115}$$

Theorem 2.6

(i) *Under a condition of*

$$A(\mathbf{p}) \times B(\mathbf{p}) < 0, \tag{2.116}$$

a standard form of the quadratic functional dynamical system in Eq. (2.109) *is*

$$\dot{x} = f(x, \mathbf{p}) = a_0 g(x)(g(x) - a). \tag{2.117}$$

(i_1) *If $a_0 dg/dx|_{x^*} > 0$, then the equilibrium of $x^* = a_1^{(i)} \in S_1$ with $g(x^*) = a_1 = 0$ is stable with $df/dx|_{x^*=a_1^{(i)}} < 0$ and the equilibrium of $x^* = a_1^{(i)} \in S_2$ with $g(x^*) = a_2 = a > 0$ is unstable with $df/dx|_{x^*=a_2^{(i)}} > 0$.*

(i_2) *If $a_0 dg/dx|_{x*} < 0$, then the equilibrium of $x^* = a_1^{(i)} \in S_1$ with $g(x^*) = a_1 = 0$ is unstable with $df/dx|_{x*=a_1^{(i)}} > 0$ and the equilibrium of $x^* = a_2^{(i)} \in S_2$ with $g(x^*) = a_2 = a > 0$ is stable with $df/dx|_{x*=a_2^{(i)}} < 0$.*

(ii) *Under a condition of*

$$A(\mathbf{p}) \times B(\mathbf{p}) > 0, \tag{2.118}$$

a standard form of the quadratic functional system in Eq. (2.109) *is*

$$\dot{x} = a_0(g(x) - a)g(x). \tag{2.119}$$

(ii_1) *If $a_0 dg/dx|_{x*} > 0$, then the equilibrium of $x^* = a_2^{(i)} \in S_2$ with $g(x^*) = a_2 = 0$ is unstable with $df/dx|_{x*=a_2^{(i)}} > 0$ and the equilibrium of $x^* = a_1^{(i)} \in S_1$ with $g(x^*) = a_1 = a < 0$ is stable with $df/dx|_{x*=a_1^{(i)}} < 0$.*

(ii_2) *If $a_0 dg/dx|_{x*} < 0$, then equilibrium of $x^* = a_2^{(i)} \in S_2$ with $g(x^*) = a_2 = 0$ is stable with $df/dx|_{x*=a_2^{(i)}} < 0$ and the equilibrium of $x^* = a_1^{(i)} \in S_1$ with $g(x^*) = a_1 = a < 0$ is unstable with $df/dx|_{x*=a_1^{(i)}} > 0$.*

(iii) *For $B(\mathbf{p}) = 0$, the corresponding standard functional form with $\Delta = 0$ is*

$$\dot{x} = f(x, \mathbf{p}) = a_0(g(x))^2. \tag{2.120}$$

(iii_1) *If $a_0(\mathbf{p}) > 0$ and $dg/dx|_{x*=a_1^{(i)}} \neq 0$, then the equilibrium of $x^* = a_1^{(i)} \in S_1$ with $g(x^*) = a_1 = 0$ is an upper-saddle of the second order with $d^2 f/dx^2|_{x*=a_1^{(i)}} > 0$. The equilibrium of $x^* = a_1^{(i)} \in S_1$ with $g(x^*) = a_1 = 0$ for two equilibriums switching is an upper-saddle-node bifurcation of the second order.*

(iii_2) *If $a_0(\mathbf{p}) < 0$ and $dg/dx|_{x*=a_1^{(i)}} \neq 0$, then the equilibrium of $x^* = a_1^{(i)} \in S_1$ with $g(x^*) = a_1 = 0$ is a lower-saddle of the second order with $d^2 f/dx^2|_{x*=a_1^{(i)}} < 0$. The equilibrium of $x^* = a_1^{(i)} \in S_1$ with $g(x^*) = a_1 = 0$ for two equilibriums switching is a lower-saddle-node bifurcation of the second order.*

(iv) *Under a condition of*

$$\Delta = a_0^2 a^2 > 0 \text{ and } dg/dx|_{x*=a_j^{(i)}} = 0, \ d^2 g/dx^2|_{x*=a_j^{(i)}} \neq 0 \ (j \in \{1, 2\}) \tag{2.121}$$

$$\{a_1, a_2\} = \text{sort}\{a, 0\}, \ a_2 > a_1;$$

a standard form of the quadratic functional dynamical system in Eq. (2.109) *is*

$$\dot{x} = f(x, \mathbf{p}) = a_0(g(x) - a_1)(g(x) - a_2). \tag{2.122}$$

(iv_1) *If $d^2 f / dx^2|_{x^*=a_j^{(i)}} > 0$, the repeated equilibrium of $x^* = a_j^{(i)} \in S_j (j \in \{1,2\})$ with $g(x^*) = a_j$ is an upper-saddle of the second order. The bifurcation at the repeated equilibrium of $x^* = a_j^{(i)} \in S_j$ for the switching of two simple equilibriums of $g(x^*) = a_j$ is called an upper-saddle-node bifurcation.*

(iv_2) *If $d^2 f / dx^2|_{x=a_j^{(i)}} < 0$, the repeated equilibrium of $x^* = a_j^{(i)} \in S_j (j \in \{1,2\})$ with $g(x^*) = a_j$ is a lower-saddle of the second order. The bifurcation at the repeated equilibrium of $x^* = a_j^{(i)} \in S_j (j = 1,2)$ for the switching of two simple equilibriums of $g(x^*) = a_j$ is called a lower-saddle-node bifurcation.*

(v) *Under a condition of*

$$\Delta = a_0^2 a^2 = 0 \text{ and } dg/dx|_{x^*=a_1^{(i)}} = 0, \ d^2 g/dx^2|_{x^*=a_1^{(i)}} \neq 0$$
$$\text{with } a_1 = a = 0,$$

(2.123)

a standard form of the quadratic functional dynamical system in Eq. (2.109) *is*

$$\dot{x} = f(x, \mathbf{p}) = a_0 (g(x))^2.$$

(2.124)

(v_1) *If $a_0 > 0$, the repeated equilibrium of $x^* = a_1^{(i)} \in S_1$ with $g(x^*) = 0$ is an upper-saddle of the fourth order. The bifurcation at the repeated equilibrium of $x^* = a_1^{(i)} \in S_1$ for the switching of two second-order upper-saddle equilibriums of $g(x^*) = a_1$ is called a fourth-order upper-saddle-node bifurcation.*

(v_2) *If $a_0 < 0$, the repeated equilibrium of $x^* = a_1^{(i)} \in S_1$ with $g(x^*) = a_1(\mathbf{p})$ is an lower-saddle of the fourth order. The bifurcation at the repeated equilibrium of $x^* = a_1^{(i)} \in S_1$ for the switching of two second-order lower-saddle equilibriums of $g(x^*) = a_1$ is called a fourth-order lower-saddle-node bifurcation.*

Proof. The theorem can be proved as for Theorem 2.4.   □

The stability and bifurcation of two equilibriums for the 1-dimensional system in Eq. (2.92) with $\Delta = B^2 - 4AC = a_0^2(a-b)^2 \geq 0$ are presented in Fig. 2.5. $a_0 = A$. The stable and unstable equilibriums varying with the vector parameter are depicted by solid and dashed curves, respectively. The bifurcation point of equilibriums occurs at the double-repeated equilibrium at the boundary of $\mathbf{p}_0 \in \partial \Omega_{12}$. With varying parameters, the two equilibriums of $x^* = a_j^{(i)} (j = 1, 2)$ with $g(x^*) = a, b$ equal each other (i.e., $g(x^*) = a = b$). Such an equilibrium is a bifurcation point at $g(x^*) = a = b$ for $\Delta = 0$. The equilibriums of $x^* = a_j^{(i)}$ $(j = 1, 2)$ with $g(x^*) = a, b$ for $\Delta \geq 0$ are presented in Figs. 2.5a,b, for $a_0 > 0$ and $a_0 < 0$ with

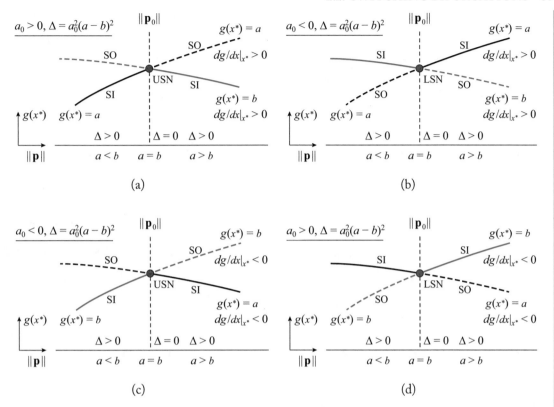

Figure 2.5: Stability and bifurcation of two equilibriums in the quadratic dynamical system: (a) an *upper-saddle*-node bifurcation ($a_0 > 0$ and $dg/dx|_{x^*} > 0$), (b) a *lower-saddle-node* bifurcation ($a_0 < 0$ and $dg/dx|_{x^*} > 0$); (c) an *upper-saddle-node* bifurcation ($a_0 > 0$ and $dg/dx|_{x^*} < 0$), (d) a *lower-saddle-node* bifurcation ($a_0 < 0$ and $dg/dx|_{x^*} < 0$). Stable and unstable equilibriums are represented by solid and dashed curves, respectively. SO: source, SI: sink.

$dg/dx|_{x^*=a_j^{(i)}} > 0$, respectively. The equilibriums of $x^* = a_j^{(i)}$ ($j = 1, 2$) with $g(x^*) = a, b$ for $\Delta \geq 0$ are presented in Figs. 2.5c,d, for $a_0 > 0$ and $a_0 < 0$ with $dg/dx|_{x^*=a_j^{(i)}} < 0$, respectively.

The dynamical system in Eq. (2.109) is as a special case of the dynamical system in Eq. (2.1) with $C(\mathbf{p}) = 0$. Thus, $\Delta = B^2 - 4AC = B^2 \geq 0$. The equilibriums exist in the entire domain. In Fig. 2.6a, for $a_0 > 0$ and $B < 0$, the equilibriums of $x^* = a_j^{(i)}$ ($j = 1, 2$) with $g(x^*) = 0$ and $g(x^*) = a$ are unstable and stable with $dg/dx|_{x^*=a_j^{(i)}} > 0$, respectively. However, in Fig. 2.6b, for $a_0 > 0$ and $B > 0$, the equilibriums of $x^* = a_j^{(i)}$ ($j = 1, 2$) with $g(x^*) = 0$ and $g(x^*) = a$ are stable and unstable with $dg/dx|_{x^*=a_j^{(i)}} > 0$, respectively. The bifurcation of functional equilibriums occurs at $B = 0$. The flow of $x(t)$ is a forward *upper* flow for $a_0 > 0$,

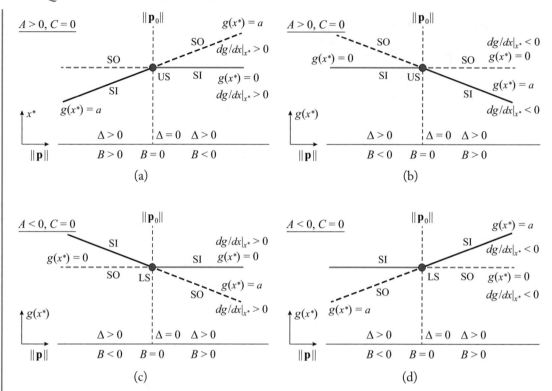

Figure 2.6: Stability and bifurcation of two equilibriums in the quadratic dynamical system: (a) an *upper-saddle-node* bifurcation ($a_0 > 0$ and $dg/dx|_{x^*} > 0$), (b) an *upper-saddle-node* bifurcation ($a_0 > 0$ and $dg/dx|_{x^*} < 0$); (c) a *lower-saddle-node* bifurcation ($a_0 < 0$ and $dg/dx|_{x^*} > 0$), (d) a *lower-saddle-node* bifurcation ($a_0 < 0$ and $dg/dx|_{x^*} < 0$). Stable and unstable equilibriums are represented by solid and dashed curves, respectively. SO: source, SI: sink.

and the equilibrium point of $x^* = a_1^{(i)} \in S_1$ with $g(x^*) = 0$ at $B = 0$ is termed an *upper-saddle*. Such a bifurcation is termed an *upper-saddle-node* bifurcation. In Fig. 2.6c, for $a_0 < 0$ and $B < 0$, the equilibriums of $x^* = a_j^{(i)} \in S_j$ ($j = 1, 2$) with $g(x^*) = 0$ and $g(x^*) = a$ are stable and unstable with $dg/dx|_{x^* = a_j^{(i)}} > 0$, respectively. However, in Fig. 2.6d, for $a_0 < 0$ and $B > 0$, the equilibriums of $x^* = a_j^{(i)} \in S_j$ ($j = 1, 2$) with $g(x^*) = 0$ and $g(x^*) = a$ are unstable and stable with $dg/dx|_{x^* = a_j^{(i)}} < 0$, respectively. The bifurcation of equilibriums also occurs at $B = 0$.

The flow of $x(t)$ is a forward *lower* flow for $a_0 < 0$, and the equilibrium point of $x^* = a_1^{(i)}$ with $g(x^*) = 0$ at $B = 0$ is termed a *lower-saddle*. Such a bifurcation is termed a *lower-saddle-node* bifurcation.

To illustrate functional equilibrium switching, due to $\Delta = a_0^2(a-b)^2 \geq 0$, consider $g(x^*) = \alpha_0(x^*)^2 + \alpha_1 x^* + \alpha_2 = a_i$ with $\Delta_i = \alpha_1^2 - 4\alpha_0(\alpha_2 - a_i)$ and $a_i = a, b$ ($i = 1, 2$). $a_0 = A$. Four equilibriums in the quadratic functional dynamical system with a quadratic function are presented in Fig. 2.7. A functional *upper-saddle-node* bifurcation ($a_0 > 0, \Delta = 0$) and a functional *lower-saddle-node* bifurcation ($a_0 < 0, \Delta = 0$) for two functional equilibrium switching are sketched in Figs. 2.7a,b, respectively. Stable and unstable equilibriums are represented by thick solid and dashed curves, respectively. The acronyms SO and SI are for source and sink equilibriums. The acronyms US and LS are for *upper-* and *lower-saddle* equilibriums, respectively. The thin curves are for the quadratic functions of $g(x^*) = a_i (i = 1, 2)$. For such a quadratic function in the quadratic functional system, there are four simple equilibriums and three double-repeated equilibriums.

To illustrate equilibrium switching in the quadratic nonlinear functional dynamical system, due to $\Delta = a_0^2 a^2 \geq 0$, consider $g(x^*) = \alpha_0(x^*)^2 + \alpha_1 x^* + \alpha_2 = a_i$ with $\Delta_i = \alpha_1^2 - 4\alpha_0(\alpha_2 - a_i)$ and $a_i = a, 0$ ($i = 1, 2$). $g(x^*) = 0$ is invariant. Four equilibriums in the quadratic functional dynamical system with a quadratic function are presented in Fig. 2.8. The functional equilibrium switching with a functional *upper-saddle-node* bifurcation ($a_0 > 0, \Delta = 0$) and a functional *lower-saddle-node* bifurcation ($a_0 < 0, \Delta = 0$) are sketched in Figs. 2.8a,b, respectively. Stable and unstable equilibriums are also depicted by thick solid and dashed curves, respectively. The acronyms SO and SI are also for source and sink equilibriums. The acronyms US and LS are also for *upper-* and *lower-saddle* equilibriums, respectively. The thin curves are for the quadratic functions of $g(x^*) = a_i$ ($i = 1, 2$). For such a quadratic function in the quadratic functional system, there are two and four simple equilibriums and four double-repeated equilibriums.

2.3   APPEARING BIFURCATIONS

Definition 2.7   If $B(\mathbf{p}) = 0$ in Eq. (2.1), a 1-dimensional quadratic nonlinear functional dynamical system is

$$\dot{x} = A(\mathbf{p})(g(x))^2 + C(\mathbf{p}). \tag{2.125}$$

(i) For $A(\mathbf{p}) \times C(\mathbf{p}) > 0$, the functional dynamical system does not have any equilibriums.

(i$_1$) The non-equilibrium flow of the functional dynamical system is called a positive flow if $A(\mathbf{p}) > 0$.

(i$_2$) The non-equilibrium flow of the functional dynamical system is called a negative flow if $A(\mathbf{p}) < 0$.

(ii) For $A(\mathbf{p}) \times C(\mathbf{p}) < 0$, the corresponding standard functional form is

$$\dot{x} = a_0(g(x) + a)(g(x) - a) \tag{2.126}$$

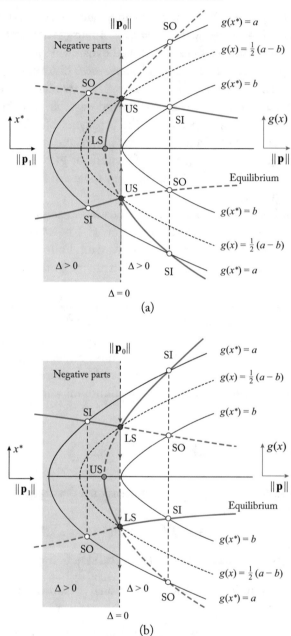

Figure 2.7: Two functional equilibrium switching in the quadratic nonlinear functional dynamical system with a quadratic function: (a) a functional *upper-saddle-node switching* bifurcation ($a_0 > 0$), (b) a functional *lower-saddle-node* switching bifurcation ($a_0 < 0$). Stable and unstable equilibriums are represented by solid and dashed curves, respectively. SO: source, SI: sink.

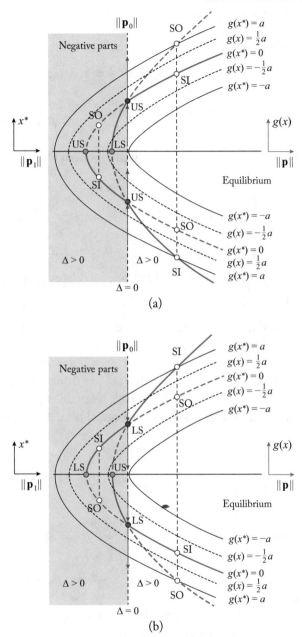

Figure 2.8: Two functional equilibrium switching in the quadratic nonlinear functional dynamical system with a quadratic function: (a) a functional *upper-saddle-node* switching bifurcation ($a_0 > 0$), (b) a functional *lower-saddle-node* switching bifurcation ($a_0 < 0$). Stable and unstable equilibriums are represented by solid and dashed curves, respectively.

with two sets of symmetric functional equilibriums as

$$S_1 = \{a_1^{(i)} | g(a_1^{(i)}) = a_1 = -a, \ i = 1, 2, \ldots, N_1\} \cup \{\emptyset\}$$

$$S_2 = \{a_2^{(i)} | g(a_2^{(i)}) = a_2 = a, \ i = 1, 2, \ldots, N_2\} \cup \{\emptyset\}$$

$$\text{with } a_0 = A(\mathbf{p_0}) \text{ and } a = \sqrt{-\frac{C(\mathbf{p_0})}{A(\mathbf{p_0})}}.$$

(2.127)

(iii) For $C(\mathbf{p_0}) = 0$, the corresponding standard functional form with $\Delta = 0$ is

$$\dot{x} = a_0[g(x)]^2$$

(2.128)

with a set of functional equilibriums as

$$S_1 = \{a_1^{(i)} | g(a_1^{(i)}) = a_1 = 0, \ i = 1, 2, \ldots, N_1\}.$$

(2.129)

Such an equilibrium point of $x^* = a_1^{(i)} \in S_1$ with $g(x^*) = 0$ is called a *saddle* of the second order. If $a_0 > 0$, the equilibrium of $x^* = a_1^{(i)} \in S_1$ with $dg/dx|_{x^*=a_1^{(i)}} \neq 0$ is an *upper-saddle* of the second order. If $a_0 < 0$, the equilibrium of $x^* = a_1^{(i)} \in S_1$ with $dg/dx|_{x^*=a_1^{(i)}} \neq 0$ is a *lower-saddle* of the second order.

(iv) The equilibrium of $x^* = a_1^{(i)} \in S_1$ with $g(x^*) = 0$ for two functional equilibriums appearance or vanishing is called a *saddle-node* bifurcation point of functional equilibrium at a point $\mathbf{p} = \mathbf{p_0} \in \partial\Omega_{12}$, and the functional bifurcation condition is

$$C(\mathbf{p_0}) = 0.$$

(2.130)

Theorem 2.8

(i) *Under a condition of*

$$A(\mathbf{p}) \times C(\mathbf{p}) < 0,$$

(2.131)

a standard functional form of the quadratic nonlinear functional dynamical system in Eq. (2.125) *is*

$$\dot{x} = f(x, \mathbf{p}) = a_0(g(x) + a)(g(x) - a).$$

(2.132)

(i_1) *If $a_0(\mathbf{p}) > 0$, then the equilibrium of $x^* = a_1^{(i)} \in S_1$ with $g(x^*) = -a$ is stable for $dg/dx|_{x^*=a_1^{(i)}} > 0$ and unstable for $dg/dx|_{x^*=a_1^{(i)}} < 0$, and the equilibrium of $x^* = a_2^{(i)} \in S_2$ with $g(x^*) = a$ is unstable for $dg/dx|_{x^*=a_2^{(i)}} > 0$ and stable for $dg/dx|_{x^*=a_2^{(i)}} < 0$.*

(i_2) If $a_0(\mathbf{p}) < 0$, then the equilibrium of $x^* = a_1^{(i)} \in S_1$ with $g(x^*) = -a$ is unstable for $dg/dx|_{x^*=a_1^{(i)}} > 0$ and stable for $dg/dx|_{x^*=a_1^{(i)}} < 0$, and the equilibrium of $x^* = a_2^{(i)} \in S_2$ with $g(x^*) = a$ is stable for $dg/dx|_{x^*=a_2^{(i)}} > 0$ and unstable for $dg/dx|_{x^*=a_2^{(i)}} < 0$.

(ii) *Under a condition of*

$$C(\mathbf{p}) = 0, \ dg/dx|_{x^*=a_1^{(i)}} \neq 0 \tag{2.133}$$

a standard functional form of the quadratic nonlinear functional dynamical system in Eq. (2.125) is

$$\dot{x} = f(x, \mathbf{p}) = a_0(g(x))^2. \tag{2.134}$$

(ii_1) If $a_0(\mathbf{p}) > 0$, then the equilibrium of $x^* = a_1^{(i)} \in S_1$ with $g(x^*) = 0$ is an upper-saddle of the second order with $d^2 f/dx^2|_{x^*=a_1^{(i)}} > 0$. Such a bifurcation for two equilibriums appearance or vanishing is an upper-saddle-node bifurcation of the second order.

(ii_2) If $a_0(\mathbf{p}) < 0$, then the equilibrium of $x^* = a_1^{(i)} \in S_1$ with $g(x^*) = 0$ is a lower-saddle of the second order with $d^2 f/dx^2|_{x^*=a_1^{(i)}} > 0$. Such a bifurcation for two equilibriums appearance or vanishing is a lower-saddle-node bifurcation of the second order.

(iii) *Under a condition of*

$$A \times C < 0, \ and \ dg/dx|_{x^* \in a_j^{(i)}} = 0, \ d^2g/dx^2|_{x^*=a_j^{(i)}} \neq 0 \tag{2.135}$$

a standard functional form of the quadratic nonlinear functional dynamical system in Eq. (2.125) is

$$\dot{x} = f(x, \mathbf{p}) = a_0(g(x) - a_1)(g(x) - a_2). \tag{2.136}$$

(iii_1) If $d^2 f/dx^2|_{x^*=a_j^{(i)}} > 0$, the repeated equilibrium of $x^* = a_j^{(i)} \in S_j$ $(j \in \{1, 2\})$ with $g(x^*) = a_j(\mathbf{p})$ is an upper-saddle of the second order. The bifurcation at the repeated equilibrium of $x^* = a_j^{(i)} \in S_j$ for the appearance or vanishing of two simple equilibriums of $g(x^*) = a_j(\mathbf{p})$ is called an upper-saddle-node bifurcation.

(iii_2) If $d^2 f/dx^2|_{x=a_j^{(i)}} < 0$, the repeated equilibrium of $x^* = a_j^{(i)} \in S_j$ $(j \in \{1, 2\})$ with $g(x^*) = a_j(\mathbf{p})$ is a lower-saddle of the second order. The bifurcation at the repeated equilibrium of $x^* = a_j^{(i)} \in S_j$ $(j = 1, 2)$ for the appearance or vanishing of two simple equilibriums of $g(x^*) = a_j(\mathbf{p})$ is called a lower-saddle-node bifurcation.

(iv) *Under a condition of*

$$C = 0 \ and \ dg/dx|_{x^* \in S_1} = 0 \tag{2.137}$$

a standard functional form of the quadratic functional dynamical system in Eq. (2.125) is

$$\dot{x} = f(x, \mathbf{p}) = a_0(g(x))^2. \tag{2.138}$$

(iv_1) If $a_0 > 0$ and $d^2g/dx^2|_{x^*=a_1^{(i)}} \neq 0$, the repeated equilibrium of $x^* = a_1^{(i)} \in S_1$ with $g(x^*) = 0$ is an upper-saddle of the fourth order. The bifurcation at the repeated equilibrium of $x^* = a_1^{(i)} \in S_1$ for the appearance or vanishing of two second-order upper-saddle functional equilibriums of $g(x^*) = 0$ is called a fourth-order upper-saddle-node bifurcation.

(iv_2) If $a_0 < 0$ and $d^2g/dx^2|_{x^*=a_1^{(i)}} \neq 0$, the repeated equilibrium of $x^* = a_1^{(i)} \in S_1$ with $g(x^*) = 0$ is an lower-saddle of the fourth order. The bifurcation at the repeated equilibrium of $x^* = a_1^{(i)} \in S_1$ for the appearance or vanishing of two second-order lower-saddle functional equilibriums of $g(x^*) = 0$ is called a fourth-order lower-saddle-node bifurcation.

Proof. The proof is similar to Theorem 2.2. The theorem is proved.   □

The stability and bifurcation of equilibriums for the quadratic nonlinear functional system in Eq. (2.56) are illustrated in Fig. 2.9 as a special case of the dynamical system in Eq. (2.1) with $B(\mathbf{p}) = 0$. The stable and unstable equilibriums varying with the vector parameter are depicted by solid and dashed curves, respectively. The bifurcation point of equilibrium occurs at the double equilibrium at the boundary of $\mathbf{p}_0 \in \partial\Omega_{12}$. In Figs. 2.9a,b, for $\Delta = -4AC > 0$ and $a_0 = A > 0$, the equilibriums of $x^* = a_j^{(i)} \in S_j (j = 1, 2)$ with $g(x^*) = -a < 0$ and $g(x^*) = a > 0$ for $C < 0$ are, respectively, stable and unstable with $dg/dx|_{x^*=a_1^{(i)}} > 0$ and unstable and stable with $dg/dx|_{x^*=a_1^{(i)}} < 0$. The bifurcation of functional equilibrium also occurs at $C = 0$. The flow of $x(t)$ is a forward *upper* flow for $a_0 > 0$ and $dg/dx|_{x^*=a_1^{(i)}} \neq 0$, and the equilibrium point of $x^* = a_1^{(i)} \in S_1$ with $g(x^*) = 0$ at $C = 0$ is termed the *upper-saddle*. Such a bifurcation is termed the *upper-saddle-node* bifurcation. For $\Delta = -4AC < 0$ and $a_0 = A > 0$, we have $C > 0$. Thus, no any equilibrium exists because of $\dot{x} = Ax^2 + C > 0$. Such a 1-dimensional system is termed a non-equilibrium system. For $a_0 = A > 0$ and $C > 0$, the flow of $x(t)$ is always toward the positive direction. In Figs. 2.9c,d, for $\Delta = -4AC > 0$ and $a_0 = A < 0$, the equilibriums of $x^* = a_j^{(i)} \in S_j (j = 1, 2)$ with $g(x^*) = -a < 0$ and $g(x^*) = a > 0$ for $C < 0$ are, respectively, unstable and stable with $dg/dx|_{x^*=a_1^{(i)}} > 0$ and stable and unstable with $dg/dx|_{x^*=a_1^{(i)}} < 0$. The bifurcation of functional equilibrium also occurs at $C = 0$. The flow of $x(t)$ for the bifurcation point is a forward *lower* flow for $a_0 = A < 0$, and the functional equilibrium bifurcation point of $x^* = a_1^{(i)} \in S_1$ with $g(x^*) = 0$ at $C = 0$ is termed a *lower-saddle*. Such a bifurcation is termed a *lower-saddle-node* bifurcation. For $\Delta = -4AC < 0$ and $a_0 = A < 0$, we have $C < 0$. For $a_0 = A < 0$ and $C < 0$, the flow of $x(t)$ is always toward the negative direction without any functional equilibrium because of $\dot{x} = A(g(x))^2 + C < 0$.

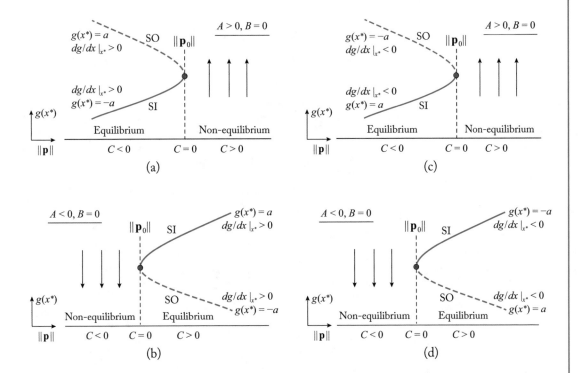

Figure 2.9: Stability and bifurcation of two equilibriums in the quadratic dynamical system: (a, b) an *upper-saddle-node* bifurcation ($a_0 > 0$), (c, d) a *lower-saddle-node* bifurcation ($a_0 < 0$). Stable and unstable equilibriums are represented by solid and dashed curves, respectively. SO: source, SI: sink.

CHAPTER 3

Cubic Nonlinear Functional Systems

In this chapter, the stability and stability switching of equilibriums in cubic polynomial functional dynamical systems are discussed. The *upper-saddle-node* and *lower-saddle-node* appearing and switching bifurcations are discussed, and the *third-order sink* and *source* switching bifurcations are discussed as well.

Definition 3.1 Consider a cubic nonlinear functional dynamical system

$$
\begin{aligned}
\dot{x} = f(x, \mathbf{p}) &= A(\mathbf{p})(g(x))^3 + B(\mathbf{p})(g(x))^2 + C(\mathbf{p})g(x) + D(\mathbf{p}) \\
&\equiv a_0(\mathbf{p})(g(x) - a(\mathbf{p}))[(g(x))^2 + B_1(\mathbf{p})g(x) + C_1(\mathbf{p})],
\end{aligned}
\tag{3.1}
$$

where four scalar constants $A(\mathbf{p}) \neq 0$, $B(\mathbf{p})$, $C(\mathbf{p})$, and $D(\mathbf{p})$ are determined by

$$
\begin{aligned}
&A = a_0, \ B = (-a + B_1)a_0, \ C = (-aB_1 + C_1)a_0, \ D = -aa_0C_1, \\
&\mathbf{p} = (p_1, p_2, \ldots, p_m)^\mathrm{T}.
\end{aligned}
\tag{3.2}
$$

(i) If

$$
\Delta_1 = B_1^2 - 4C_1 < 0 \text{ for } \mathbf{p} \in \Omega_1 \subset \mathbf{R}^m,
\tag{3.3}
$$

then the cubic nonlinear system has an equilibrium set as

$$
S_1 = \{a_1^{(i)} | g(a_1^{(i)}) = a_1 = a, \ i = 1, 2, 3, \ldots, N_1\} \cup \{\emptyset\}
\tag{3.4}
$$

and the standard functional form of such a quadratic nonlinear functional dynamical system is

$$
\dot{x} = a_0(g(x) - a)[(g(x) + \frac{1}{2}B_1)^2 + \frac{1}{4}(-\Delta_1)].
\tag{3.5}
$$

(ii) If

$$
\Delta_1 = B_1^2 - 4C_1 > 0 \text{ for } \mathbf{p} \in \Omega_2 \subset \mathbf{R}^m,
\tag{3.6}
$$

then there are three functional equilibriums with

$$
\begin{aligned}
&a_0 = A(\mathbf{p}), \ b_{1,2} = -\frac{1}{2}(B_1(\mathbf{p}) \mp \sqrt{\Delta_1}) \text{ with } b_1 > b_2; \\
&a_1 = \min\{a, b_1, b_2\}, \ a_3 = \max\{a, b_1, b_2\}, \ a_2 \in \{a, b_1, b_2\}, a_2 \notin \{a_1, a_3\}, \\
&\Delta_{ij} = (a_i - a_j)^2 > 0 \text{ for } i, j \in \{1, 2, 3\} \text{ but } i \neq j
\end{aligned}
\tag{3.7}
$$

and three functional equilibrium sets are defined as

$$S_\alpha = \{a_\alpha^{(i_\alpha)} | g(a_\alpha^{(i_\alpha)}) = a_\alpha, i_\alpha = 1, 2, 3, \ldots, N_\alpha\} \cup \{\emptyset\}, \ \alpha = 1, 2, 3. \tag{3.8}$$

(ii$_1$) If

$$a_i \neq a_j \text{ with } \Delta_{ij} = (a_i - a_j)^2 > 0 \text{ for } i, j \in \{1, 2, 3\} \text{ but } i \neq j. \tag{3.9}$$

the cubic nonlinear functional system has three functional equilibriums as

$$g(x^*) = a_\alpha, \ \alpha = 1, 2, 3, \tag{3.10}$$

and the corresponding standard functional form is

$$\dot{x} = a_0(g(x) - a_1)(g(x) - a_2)(g(x) - a_3). \tag{3.11}$$

(ii$_2$) If at $\mathbf{p} = \mathbf{p}_1$

$$\begin{aligned} a_1 &= b_2, \ a_2 = a, \ a_3 = b_1; \\ \Delta_{12} &= (a_1 - a_2) = (a - b_2)^2 = 0, \end{aligned} \tag{3.12}$$

the cubic nonlinear functional system has a double-repeated functional equilibrium and a simple functional equilibrium as

$$\begin{aligned} S_1 &= \{a_1^{(i)} | g(x^*) = a_1 = a, \ i = 1, 2, \ldots, N_1\} \\ S_2 &= \{a_2^{(i)} | g(x^*) = a_2 = b_1, \ i = 1, 2, \ldots, N_2\} \\ &\text{with } g(x^*) = a_1 = a_3 = a < b_1, \end{aligned} \tag{3.13}$$

and the corresponding standard functional form is

$$\dot{x} = a_0(g(x) - a_1)^2(g(x) - a_2). \tag{3.14}$$

Such a flow at the equilibrium of $x^* = a_1^{(i)} \in S_1$ with $g(x^*) = a_1$ for $dg/dx|_{x^*=a_1^{(i)}} \neq 0$ is called a *saddle* flow of the second order.

The equilibrium of $x^* = a_1^{(i)}$ with $g(x^*) = a_1$ for two different functional equilibriums switching is called a bifurcation point of functional equilibriums at a point $\mathbf{p} = \mathbf{p}_1$ with the second-order multiplicity, and the functional bifurcation condition is

$$a = b = \min\{-\frac{1}{2}(B_1(\mathbf{p}) + \sqrt{\Delta_1}), \ -\frac{1}{2}(B_1(\mathbf{p}) - \sqrt{\Delta_1})\}. \tag{3.15}$$

(ii$_3$) If at $\mathbf{p} = \mathbf{p}_2$,

$$a_2 = b_1, \ a_3 = a, \ a_1 = b_2,$$
$$\Delta_{23} = (a_2 - a_3) = (a - b_1)^2 = 0, \tag{3.16}$$

the cubic nonlinear functional system has two functional equilibriums as

$$S_1 = \{a_1^{(i)} | g(x^*) = a_1 = b_2, \ i = 1, 2, \ldots, N_1\}$$
$$S_2 = \{a_2^{(i)} | g(x^*) = a_2 = a, \ i = 1, 2, \ldots, N_2\} \tag{3.17}$$
$$\text{with } g(x^*) = a_2 = a_3 = a > b_2,$$

and the corresponding standard functional form is

$$\dot{x} = a_0(g(x) - a_1)(g(x) - a_2)^2. \tag{3.18}$$

Such a flow at the equilibrium of $x^* = a_2^{(i)} \in S_2$ with $g(x^*) = a_2$ for $dg/dx|_{x^*=a_2^{(i)}} \neq 0$ is called a *saddle* flow of the second-order.

The equilibrium of $x^* = a_2^{(i)} \in S_2$ with $g(x^*) = a_2$ for two different functional equilibriums switching is called a bifurcation point of equilibrium at a point $\mathbf{p} = \mathbf{p}_2$ with the second-order multiplicity, and the bifurcation condition is

$$a = b_2 = \max\{-\frac{1}{2}(B_1(\mathbf{p}) + \sqrt{\Delta_1}), \ -\frac{1}{2}(B_1(\mathbf{p}) - \sqrt{\Delta_1})\}. \tag{3.19}$$

(ii$_4$) If at $\mathbf{p} = \mathbf{p}_3$,

$$a_1 = b_2, \ a_2 = a, \ a_3 = b_1,$$
$$\Delta_{12} = (a_1 - a_2)^2 = (a - b_2)^2 = 0,$$
$$\Delta_{23} = (a_2 - a_3)^2 = (a - b_1)^2 = 0, \tag{3.20}$$
$$\Delta_{13} = (a_1 - a_3)^2 = (b_2 - b_1)^2 = 0,$$

the cubic nonlinear functional system has three repeated functional equilibriums as

$$S_1 = \{a_1^{(i)} | g(x^*) = a_1 = a, \ i = 1, 2, \ldots, N_1\}$$
$$g(x^*) = a_1 = a_2 = a_3 = a \tag{3.21}$$

and the corresponding standard functional form is

$$\dot{x} = a_0(g(x) - a)^3. \tag{3.22}$$

Such a flow at the equilibrium of $x^* = a_1^{(i)} \in S_1$ with $g(x^*) = a_1 = a$ for $dg/dx|_{x^*=a_1^{(i)}} \neq 0$ is called a *sink* or *source* flow of the third order.

The equilibrium of $x^* = a_1^{(i)} \in S_1$ with $g(x^*) = a_1 = a$ at a point $\mathbf{p} = \mathbf{p}_3$ for three different functional equilibriums switching is called a bifurcation point of functional equilibrium with the third-order multiplicity, and the functional bifurcation condition is

$$a = b = -\frac{1}{2} B_1(\mathbf{p}). \tag{3.23}$$

(iii) If

$$\Delta_1 = B_1^2 - 4A_1 C_1 = 0 \text{ for } \mathbf{p} = \mathbf{p}_0 \in \partial\Omega_{12} \subset \mathbf{R}^{m-1}, \tag{3.24}$$

then there exist

$$a_0 = A(\mathbf{p}_0), \text{ and } b_1 = b_2 = b = -\frac{1}{2} B_1(\mathbf{p}_0). \tag{3.25}$$

(iii$_1$) For $a < b$, the cubic nonlinear functional dynamical system has an *upper* double-repeated functional equilibrium plus a *lower* simple functional equilibrium

$$S_1 = \{a_1^{(i)} | g(x^*) = a_1 = a, \ i = 1, 2, \dots, N_1\}$$
$$S_2 = \{a_2^{(i)} | g(x^*) = a_2 = b, \ i = 1, 2, \dots, N_2\} \tag{3.26}$$
$$\text{with } g(x^*) = a_2 = a_3 = b > a$$

with the corresponding standard functional form of

$$\dot{x} = a_0 (g(x) - a_1)(g(x) - a_2)^2. \tag{3.27}$$

Such a flow at the functional equilibrium of $x^* = a_2^{(i)} \in S_2$ with $g(x^*) = a_2$ is called a *saddle* flow of the second order.

The equilibrium of $x^* = a_2^{(i)} \in S_2$ with $g(x^*) = a_2$ for two different functional equilibrium switching is called a bifurcation point of functional equilibrium at a point $\mathbf{p} = \mathbf{p}_0 \in \partial\Omega_{12}$ with the second-order multiplicity, and the bifurcation condition is

$$\Delta_1 = B_1^2 - 4C_1 = 0 \text{ with } a < b. \tag{3.28}$$

(iii$_2$) For $a > b$, the cubic nonlinear functional system has a set of *lower* double-repeated functional equilibrium plus a set of *upper* simple functional equilibrium as

$$S_1 = \{a_1^{(i)} | g(x^*) = a_1 = b, \ i = 1, 2, \dots, N_1\}$$
$$S_2 = \{a_2^{(i)} | g(x^*) = a_2 = a, \ i = 1, 2, \dots, N_2\} \tag{3.29}$$
$$\text{with } g(x^*) = a_1 = a_3 = b < a$$

with the corresponding standard form of

$$\dot{x} = a_0 (x - a_1)^2 (x - a_2). \tag{3.30}$$

Such a flow at the functional equilibrium of $x^* = a_1^{(i)} \in S_1$ with $g(x^*) = a_1 = b$ is called a *saddle* flow of the second order.

The equilibrium of $x^* = a_1^{(i)} \in S_1$ with $g(x^*) = a_1 = b$ for two different functional equilibriums switching is called a bifurcation point of functional equilibrium at a point $\mathbf{p} = \mathbf{p_0} \in \partial\Omega_{12}$ with the second-order multiplicity, and the bifurcation condition is also

$$\Delta_1 = B_1^2 - 4C_1 = 0 \text{ with } a > b. \tag{3.31}$$

(iii$_3$) For $a = b$, the cubic nonlinear functional system has a set of triple-repeated equilibrium as

$$S_1 = \{a_1^{(i)}|g(x^*) = a_1 = b, \ i = 1, 2, \ldots, N_1\}$$
$$\text{with } g(x^*) = a_1 = a_2 = a_3 = a \tag{3.32}$$

with the corresponding standard functional form of

$$\dot{x} = a_0(g(x) - a_1)^3. \tag{3.33}$$

Such a flow at the equilibrium of $x^* = a_1^{(i)} \in S_1$ with $g(x^*) = a_1 = a$ is called a *source* or *sink* flow of the third order.

The equilibrium of $x^* = a_1^{(i)} \in S_1$ with $g(x^*) = a_1 = a$ for three equilibriums switching or two equilibriums switching is called a bifurcation point of equilibrium at a point $\mathbf{p} = \mathbf{p_0} \in \partial\Omega_{12}$ with the third-order multiplicity, and the bifurcation condition is

$$\Delta_1 = B_1^2 - 4C_1 = 0 \text{ with } a = b. \tag{3.34}$$

Theorem 3.2

(i) *Under a condition of*

$$\Delta_1 = B_1^2 - 4C_1 < 0 \tag{3.35}$$

a standard form of the cubic nonlinear functional dynamical system in Eq. (3.1) *is*

$$\dot{x} = f(x, \mathbf{p}) = a_0(g(x) - a_1)[(g(x) + \frac{1}{2}B_1)^2 + \frac{1}{4}(-\Delta_1)]. \tag{3.36}$$

(i$_1$) *For $a_0 dg/dx|_{x^*} > 0$, the equilibrium of $x^* = a_1^{(i)} \in S_1$ with $g(x^*) = a_1$ is unstable with $df/dx|_{x^*=a_1} > 0$.*

(i$_2$) *For $a_0 dg/dx|_{x^*} < 0$, the equilibrium of $x^* = a_1^{(i)} \in S_1$ with $g(x^*) = a_1$ is stable with $df/dx|_{x^*=a_1} < 0$.*

(ii) Under the conditions of

$$\Delta_1 = B_1^2 - 4C_1 > 0,$$
$$a_1, a_2, a_3 = sort\{b_2, a, b_1\}, \ a_i \neq a_j, \ a_i < a_{i+1}; \tag{3.37}$$
$$\Delta_{ij} = (a_i - a_j)^2 \neq 0 \ for \ i, j, \in \{1, 2, 3\},$$

a standard functional form of the cubic nonlinear functional dynamical system in Eq. (3.1) *is*

$$\dot{x} = f(x, \mathbf{p}) = a_0(g(x) - a_1)(g(x) - a_2)(g(x) - a_3) \tag{3.38}$$

with

$$\frac{df}{dx}\Big|_{x^* = a_{i_1}^{(i)}} = a_0 \frac{dg}{dx}\Big|_{x^* = a_{i_1}^{(i)}} \prod_{i_2 \neq i_1}^{3} (a_{i_1} - a_{i_2}). \tag{3.39}$$

For $dg/dx|_{x^* = a_j^{(i)}} \neq 0$, the equilibriums of $x^* = a_j^{(i)} \in S_j$ with $x^* = a_j \ (j = 1, 2, 3)$ are unstable if $df/dx|_{x^* = a_j^{(i)}} > 0$ and stable if $df/dx|_{x^* = a_j^{(i)}} < 0$.

(iii) Under a condition of

$$\Delta_1 = B_1^2 - 4C_1 > 0,$$
$$a_1, a_2, a_3 = sort\{b_2, a, b_1\}, \ a_i \neq a_j, \ a_i \leq a_{i+1} \tag{3.40}$$
$$\Delta_{12} = (a_1 - a_2)^2 = 0 \ for \ i, j, \in \{1, 2, 3\},$$

a standard functional form of the cubic nonlinear functional dynamical system in Eq. (3.1) *is*

$$\dot{x} = f(x, \mathbf{p}) = a_0(g(x) - a_1)^2(g(x) - a_3). \tag{3.41}$$

(iii$_1$) The equilibriums of $x^* = a_1^{(i)} \in S_1$ with $g(x^*) = a_1$ and $x^* = a_3^{(i)} \in S_3$ with $g(x^*) = a_3$ are unstable (lower-saddle, $d^2 f/dx^2|_{x^* = a_1^{(i)}} < 0$) and unstable (source, $df/dx|_{x^* = a_3^{(i)}} > 0$), respectively. The bifurcation of equilibrium at $x^* = a_1^{(i)} \in S_1$ for the two different functional equilibriums switching is a lower-saddle-node bifurcation of the second order at a point $\mathbf{p} = \mathbf{p}_1$.

(iii$_2$) The equilibriums of $x^* = a_1^{(i)} \in S_1$ with $g(x^*) = a_1$ and $x^* = a_3^{(i)} \in S_3$ with $g(x^*) = a_3$ are unstable (upper-saddle, $d^2 f/dx^2|_{x^* = a_1^{(i)}} > 0$) and unstable (sink, $df/dx|_{x^* = a_3^{(i)}} < 0$), respectively. The bifurcation of equilibrium at $x^* = a_1^{(i)} \in S_1$ for the two different functional equilibriums switching is a lower-saddle-node bifurcation of the second order at a point $\mathbf{p} = \mathbf{p}_1$.

(iv) For

$$\Delta_1 = B_1^2 - 4C_1 > 0,$$
$$a_1, a_2, a_3 = sort\{b_2, a, b_1\}, \ a_i \neq a_j, \ a_i \leq a_{i+1} \qquad (3.42)$$
$$\Delta_{23} = (a_2 - a_3)^2 = 0 \ for \ i, j, \in \{1, 2, 3\},$$

a standard form of the 1-dimensional dynamical system in Eq. (3.1) *is*

$$\dot{x} = f(x, \mathbf{p}) = a_0(g(x) - a_1)(g(x) - a_2)^2. \qquad (3.43)$$

(iv$_1$) The equilibriums of $x^* = a_1^{(i)} \in S_1$ with $g(x^*) = a_1$ and $x^* = a_2^{(i)} \in S_2$ with $g(x^*) = a_2$ are unstable (source, $df/dx|_{x^*=a_1^{(i)}} > 0$) and unstable (upper-saddle, $d^2 f/dx^2|_{x^*=a_2^{(i)}} > 0$), respectively. The bifurcation of equilibrium at $x^* = a_2^{(i)} \in S_2$ for two different functional equilibriums switching is an upper-saddle-node bifurcation of the second order at a point $\mathbf{p} = \mathbf{p}_1$.

(iv$_2$) The equilibriums of $x^* = a_1^{(i)} \in S_1$ with $g(x^*) = a_1$ and $x^* = a_2^{(i)} \in S_2$ with $g(x^*) = a_2$ are stable (sink, $df/dx|_{x^*=a_1^{(i)}} < 0$) and unstable (lower-saddle, $d^2 f/dx^2|_{x^*=a_2^{(i)}} > 0$), respectively. The bifurcation of equilibrium at $x^* = a_2^{(i)} \in S_2$ for two functional equilibrium switching is a lower-saddle-node bifurcation of the second order at a point $\mathbf{p} = \mathbf{p}_1$.

(v) For

$$\Delta_1 = B_1^2 - 4C_1 > 0, \ b_1 = b_2$$
$$a_1, a_2, a_3 = sort\{b_2, a, b_1\}, \ a_i \leq a_{i+1}, \qquad (3.44)$$
$$\Delta_{ij} = (a_i - a_j)^2 = 0 \ for \ i, j, = 1, 2, 3 \ but \ i \neq j,$$

a standard functional form of the cubic nonlinear functional dynamical system in Eq. (3.1) *is*

$$\dot{x} = f(x, \mathbf{p}) = a_0(g(x) - a_1)^3. \qquad (3.45)$$

(v$_1$) The equilibrium of $x^* = a_1^{(i)} \in S_1$ with $g(x^*) = a_1$ is unstable (third-order source, $d^3 f/dx^3|_{x^*=a_1^{(i)}} > 0$). The bifurcation of equilibrium at $x^* = a_1^{(i)} \in S_1$ for three different functional equilibrium switching is a source switching bifurcation of the third order at a point $\mathbf{p} = \mathbf{p}_1$.

(v$_2$) The equilibrium of $x^* = a_1^{(i)} \in S_1$ with $g(x^*) = a_1$ is stable (third-order sink, $d^3 f/dx^3|_{x^*=a_1^{(i)}} < 0$). The bifurcation of equilibrium at $x^* = a_1^{(i)} \in S_1$ with $g(x^*) = a_1$ for three different functional equilibrium switching is a sink switching bifurcation of the third order at a point $\mathbf{p} = \mathbf{p}_1$.

(vi) For

$$\Delta_1 = B_1^2 - 4A_1C_1 = 0, \ a < b$$
$$a_1 = a, \ a_2 = b, \ \Delta_{12} = (a_1 - a_2)^2 \neq 0 \tag{3.46}$$

at $\mathbf{p} = \mathbf{p}_0 \in \partial\Omega_{12}$, *a standard form of the 1-dimensional dynamical system is*

$$\dot{x} = f(x, \mathbf{p}) = a_0(g(x) - a_1)(g(x) - a_2)^2. \tag{3.47}$$

(vi$_1$) The equilibriums of $x^* = a_1^{(i)} \in S_1$ *with* $g(x^*) = a_1$ *and* $x^* = a_2^{(i)} \in S_2$ *with* $g(x^*) = a_2$ *are unstable (source,* $df/dx|_{x^* = a_1^{(i)}} > 0$*) and unstable (upper-saddle,* $d^2 f/dx^2|_{x^* = a_2^{(i)}} > 0$*), respectively. The bifurcation of equilibrium at* $x^* = a_2^{(i)} \in S_2$ *for two different functional equilibriums vanishing or appearance is an upper-saddle-node bifurcation of the second order at a point* $\mathbf{p} = \mathbf{p}_0 \in \partial\Omega_{12}$.

(vi$_2$) The equilibriums of $x^* = a_1^{(i)} \in S_1$ *with* $g(x^*) = a_1$ *and* $x^* = a_2^{(i)} \in S_2$ *with* $g(x^*) = a_2$ *are stable (sink,* $df/dx|_{x^* = a_1^{(i)}} < 0$*) and unstable (lower-saddle,* $d^2 f/dx^2|_{x^* = a_2^{(i)}} < 0$*), respectively. The bifurcation of equilibrium at* $x^* = a_2$ *for two different functional equilibriums vanishing or appearance is a lower-saddle-node bifurcation of the second order at a point* $\mathbf{p} = \mathbf{p}_0 \in \partial\Omega_{12}$.

(vii) For

$$\Delta_1 = B_1^2 - 4A_1C_1 = 0, \ a > b$$
$$a_1 = b, \ a_2 = a, \ \Delta_{12} = (a_1 - a_2)^2 \neq 0 \tag{3.48}$$

at $\mathbf{p} = \mathbf{p}_0 \in \partial\Omega_{12}$, *a standard form of the 1-dimensional dynamical system is*

$$\dot{x} = f(x, \mathbf{p}) = a_0(g(x) - a_1)^2(g(x) - a_2). \tag{3.49}$$

(vii$_1$) The equilibriums of $x^* = a_1^{(i)} \in S_1$ *with* $g(x^*) = a_1$ *and* $x^* = a_2^{(i)} \in S_2$ *are unstable (lower-saddle,* $d^2 f/dx^2|_{x^* = a_1^{(i)}} < 0$*) and unstable (source,* $df/dx|_{x^* = a_2^{(i)}} > 0$*), respectively. The bifurcation of equilibrium at* $x^* = a_1^{(i)} \in S_1$ *for two different functional equilibriums appearance or vanishing is a lower-saddle-node bifurcation of the second order at a point* $\mathbf{p} = \mathbf{p}_0 \in \partial\Omega_{12}$.

(vii$_2$) The equilibriums of $x^* = a_1^{(i)} \in S_1$ *with* $g(x^*) = a_1$ *and* $x^* = a_2^{(i)} \in S_2$ *are unstable (upper-saddle,* $d^2 f/dx^2|_{x^* = a_1^{(i)}} > 0$*) and stable (sink,* $df/dx|_{x^* = a_2^{(i)}} > 0$*), respectively. The bifurcation of equilibrium at* $x^* = a_1^{(i)} \in S_1$ *for two different functional equilibrium appearance or vanishing is a upper-saddle-node bifurcation of the second order at a point* $\mathbf{p} = \mathbf{p}_0 \in \partial\Omega_{12}$.

(viii) For

$$\Delta_1 = B_1^2 - 4A_1C_1 = 0, \ a = b$$
$$a_2 = a, \ a_2 = a_3 = b,$$
$$\Delta_{12} = (a_1 - a_2)^2 = 0$$

(3.50)

at $\mathbf{p} = \mathbf{p}_0 \in \partial\Omega_{12}$, a standard functional form of the cubic nonlinear functional dynamical system is

$$\dot{x} = f(x, \mathbf{p}) = a_0(g(x) - a_1)^3.$$

(3.51)

(viii$_1$) The equilibrium $x^* = a_1^{(i)} \in S_1$ with $g(x^*) = a_1$ is unstable (third-order source, $d^3f/dx^3|_{x^*=a_1^{(i)}} > 0$). The bifurcation of equilibrium at $x^* = a_1^{(i)} \in S_1$ from one to three different functional equilibrium switching is a source switching-appearing bifurcation of the third order at a point $\mathbf{p} = \mathbf{p}_1$.

(viii$_2$) The equilibrium $x^* = a_1^{(i)} \in S_1$ with $g(x^*) = a_1$ is unstable (third-order sink, $d^3f/dx^3|_{x^*=a_1^{(i)}} < 0$). The bifurcation of equilibrium at $x^* = a_1^{(i)} \in S_1$ for one to three different functional equilibrium switching is a sink switching-appearing bifurcation of the third order at a point $\mathbf{p} = \mathbf{p}_1$.

Proof. The proof is similar to Theorem 2.2.   □

The 1-dimensional cubic nonlinear system can be expressed by a factor of $(g(x) - a)$ and a quadratic form of $a_0((g(x))^2 + B_1g(x) + C_1)$ as in Eq. (3.1). Three functional equilibriums do not have any intersections. Thus, only one bifurcation occurs at $\Delta_1 = B_1^2 - 4C_1 = 0$. The bifurcation of functional equilibrium occurs at the double-repeated equilibrium at the boundary of $\mathbf{p}_0 \in \partial\Omega_{12}$. For $\Delta_1 = B_1^2 - 4C_1 > 0$, $(g(x))^2 + B_1g(x) + C_1 = 0$ gives two functional equilibrium sets of $x^* = a_j^{(i)} \in S_j$ $(j = 1, 2)$ with $g(x^*) = b_1, b_2$. For $a_0dg/dx|_{x^*=a_j^{(i)}} > 0$ $(j = 1, 2, 3)$, if $a > \max\{b_1, b_2\}$, then the equilibrium of $x^* = a_3^{(i)} \in S_3$ with $g(x^*) = a_3 = a$ is unstable, and the equilibrium of $x^* = a_2^{(i)} \in S_2$ with $g(x^*) = a_2 = \max\{b_1, b_2\}$ and the equilibrium of $x^* = a_2^{(i)} \in S_2$ with $g(x^*) = a_1 = \min\{b_1, b_2\}$ are stable and unstable, respectively. For $\Delta_1 = B_1^2 - 4C_1 < 0$, $(g(x))^2 + B_1g(x) + C_1 = 0$ does not have any real solutions. For $\Delta_1 = B_1^2 - 4C_1 = 0$, $(g(x))^2 + B_1g(x) + C_1 = 0$ has a double-repeated functional equilibrium of $g(x^*) = b = -\frac{1}{2}B_1$. The condition of $\Delta_1 = B_1^2 - 4C_1 = 0$ gives

$$B_1^2 = 4C_1.$$

(3.52)

From Eq. (3.2), one obtains

$$B_1 = a + \frac{B}{A} \text{ and } C_1 = \frac{C}{A} + a(a + \frac{B}{A}).$$

(3.53)

Thus, Equation (3.52) gives

$$a = -\frac{B}{3A} \pm \frac{2}{3A}\sqrt{B^2 - 3AC}. \tag{3.54}$$

Further, the double-repeated functional equilibrium of $g(x^*) = b = -\frac{1}{2}B_1$ is given by

$$g(x^*) = b = -\frac{B}{3A} \mp \frac{1}{3A}\sqrt{B^2 - 3AC}. \tag{3.55}$$

If $B^2 > 3AC$, such a double-repeated functional equilibrium exists. If $B^2 < 3AC$, such a double-repeated functional equilibrium does not exist. From Eq. (3.55), another functional equilibrium is $g(x^*) = a$, which is different from $g(x^*) = b$. If $B^2 = 3AC$, such a double-repeated functional equilibrium with the functional equilibrium of $g(x^*) = a$ has an intersected point at $g(x^*) = -\frac{B}{3A}$.

The bifurcation diagram for $a > \max\{b_1, b_2\}$ with $a_0 > 0$ and $dg/dx|_{x^*} > 0$ is presented in Fig. 3.1a. The stable and unstable equilibriums varying with the vector parameter are presented by solid and dashed curves, respectively. Such a functional equilibrium of $g(x^*) = b$ is a *lower-saddle-node* (LSN) bifurcation. The equilibrium of $x^* = a_3^{(i)} \in S_3$ with $g(x^*) = a_3 = a$ is a source. The equilibrium of $x^* = a_2^{(i)} \in S_2$ with $g(x^*) = a_2 = \max\{b_1, b_2\}$ is a sink. The equilibrium of $x^* = a_1^{(i)} \in S_1$ with $g(x^*) = a_1 = \min\{b_1, b_2\}$ is a source. However, the bifurcation diagram for $a > \max\{b_1, b_2\}$ and $a_0 < 0$ and $dg/dx|_{x^*} < 0$ is presented in Fig. 3.1b. The functional equilibrium of $g(x^*) = b$ is an *upper-saddle-node* (USN) bifurcation. The equilibrium of $x^* = a_3^{(i)} \in S_3$ with $x^* = a_3 = a$ is a sink. The equilibrium of $x^* = a_2^{(i)} \in S_2$ with $g(x^*) = a_2 = \max\{b_1, b_2\}$ is a source. The equilibrium of $x^* = a_1^{(i)} \in S_1$ with $g(x^*) = a_1 = \min\{b_1, b_2\}$ is a sink. The bifurcation diagram for $a < \min\{b_1, b_2\}$ with $a_0 < 0$ and $dg/dx|_{x^*} > 0$ is presented in Fig. 3.1c. The functional equilibrium of $g(x^*) = b$ is an *upper-saddle-node* (USN) bifurcation. The equilibrium of $x^* = a_3^{(i)} \in S_3$ with $g(x^*) = a_3 = \max\{b_1, b_2\}$ is a source. The equilibrium of $x^* = a_2^{(i)} \in S_2$ with $g(x^*) = a_2 = \min\{b_1, b_2\}$ is a sink. The equilibrium of $x^* = a_1^{(i)} \in S_1$ with $g(x^*) = a_1 = a$ is a source. The bifurcation diagram for $a < \min\{b_1, b_2\}$ with $a_0 > 0$ and $dg/dx|_{x^*} < 0$ is presented in Fig. 3.1d. The functional equilibrium of $g(x^*) = b$ is a *lower-saddle-node* (LSN) bifurcation. The equilibrium of $x^* = a_3^{(i)} \in S_3$ with $g(x^*) = a_3 = \max\{b_1, b_2\}$ is a sink. The equilibrium of $x^* = a_2^{(i)} \in S_2$ with $g(x^*) = a_2 = \min\{b_1, b_2\}$ is a source. The equilibrium of $x^* = a_3^{(i)} \in S_3$ with $g(x^*) = a_1 = a$ is a sink.

For $\Delta_1 = B_1^2 - 4C_1 \geq 0$, the 1-dimensional cubic functional nonlinear system in Eq. (3.1) have three functional equilibriums. Three functional equilibriums are $g(x^*) = a, b_1, b_2$. Assume $a_i \leq a_{i+1}$ for $i = 1, 2$ with $a_{1,2,3} = \text{sort}\{a, b_1, b_2\}$. With varying parameters, two of three functional equilibriums (i.e., $a_i = a_j$ for $i, j \in \{1, 2, 3\}$ but $i \neq j$) will be intersected each other with the corresponding discriminant of $\Delta_{ij} = (a_i - a_j)^2 = 0$, and in the vicinity of the intersection point, $\Delta_{ij} = (a_i - a_j)^2 > 0$. The two intersected points of $a = b_{1,2}$

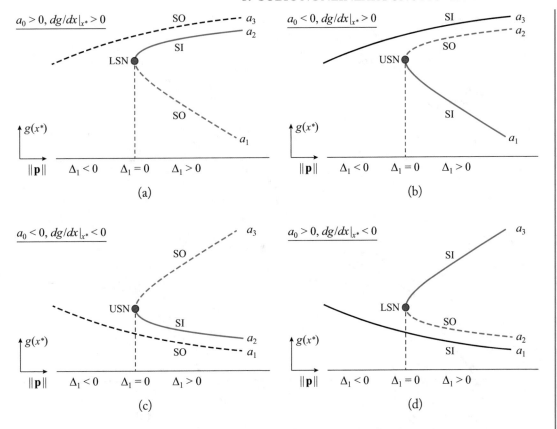

Figure 3.1: Stability and bifurcation of three independent equilibriums in the 1-dimensional, cubic nonlinear functional dynamical system. For $a > \{b_1, b_2\}$: (a) a functional LSN bifurcation ($a_0 > 0, dg/dx|_{x^*} > 0$), (b) a functional USN bifurcation ($a_0 < 0, dg/dx|_{x^*} > 0$). For $a < \{b_1, b_2\}$: (c) a functional USN bifurcation ($a_0 < 0, dg/dx|_{x^*} < 0$), (d) a functional LSN bifurcation ($a_0 > 0, dg/dx|_{x^*} < 0$). LSN: *lower-saddle-node*, USN: *upper-saddle-node*. Stable and unstable equilibriums are represented by solid and dashed curves, respectively. The bifurcation points are marked by circular symbols. SO: source, SI: sink.

gives

$$a = -\frac{B_1}{2} \pm \frac{1}{2}\sqrt{B_1^2 - 4C_1}, \tag{3.56}$$

or

$$a^2 + aB_1 = -C_1. \tag{3.57}$$

With Eq. (3.2) or Eq. (3.53), the foregoing equation gives

$$a = b_{1,2} = -\frac{B}{3A} \pm \frac{1}{3A}\sqrt{B^2 - 3AC}. \tag{3.58}$$

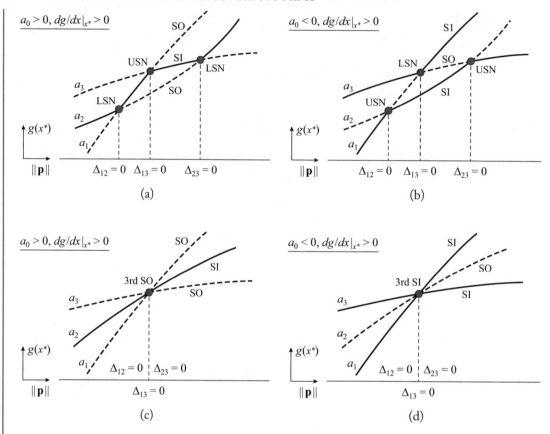

Figure 3.2: Stability and bifurcation of functional equilibriums switching in the 1-dimensional, cubic nonlinear dynamical system. For two functional equilibriums switching: (a) $a_0 > 0$ and $dg/dx|_{x^*} > 0$, (b) $a_0 < 0$ and $dg/dx|_{x^*} > 0$. For three functional equilibrium switching: (c) third-order source functional bifurcation ($a_0 > 0$ and $dg/dx|_{x^*} > 0$), (d) third-order functional sink bifurcation ($a_0 < 0$ and $dg/dx|_{x^*} > 0$). LSN: *lower–saddle-node*, USN: *upper–saddle-node*. Stable and unstable equilibriums are represented by solid and dashed curves, respectively. The bifurcation points are marked by circular symbols. SO: source, SI: sink.

If $B^2 > 3AC$, such a functional intersected point of $g(x^*) = a$ and $g(x^*) = b_1$ or $g(x^*) = b_2$ exists. If $B^2 < 3AC$, such a functional intersected point does not exist. Such the intersection point is for the two functional equilibrium switching, which is called the *saddle-node functional bifurcation*. The stability and bifurcation diagrams for $a_0 > 0$ and $a_0 < 0$ with $dg/dx|_{x^*} > 0$ are presented in Figs. 3.2a,b, respectively. Three functional equilibriums are intersected at a point with $\Delta_{ij} = (a_i - a_j)^2 = 0$ and $a_1 = a_2 = a_3 = -\frac{B}{3A}$, and in the vicinity of the intersection point, $\Delta_{ij} = (a_i - a_j)^2 > 0$ for $i, j = 1, 2, 3$ but $i \neq j$. The intersection points for $a_0 > 0$ and

$a_0 < 0$ with $dg/dx|_{x^*} > 0$ are called the source and sink functional bifurcations of the third-order, respectively. The corresponding stability and bifurcation diagrams for three equilibriums switching are presented in Figs. 3.2c,d.

In the 1-dimensional cubic nonlinear functional dynamical system of Eq. (3.1), $(g(x))^2 + B_1 g(x) + C_1 = 0$ gives two functional equilibriums of $g(x^*) = b_1, b_2$ for $\Delta_1 = B_1^2 - 4C_1 > 0$. One of the two functional equilibriums has one intersection with $g(x^*) = a$ and there are three different functional equilibriums for $a = a_2 \in (\min\{b_1, b_2\}, \max\{b_1, b_2\})$. For this case, the intersection point occurs at $a = \min\{b_1, b_2\}$ for $\mathbf{p}_1 \in \partial\Omega_{23}$ or $a = \max\{b_1, b_2\}$ for $\mathbf{p}_2 \in \partial\Omega_{23}$. The bifurcation point of functional equilibrium occurs at the double-repeated functional equilibrium at $\Delta_1 = B_1^2 - 4C_1 = 0$ for $\mathbf{p}_0 \in \partial\Omega_{12}$. Such a bifurcation is a *lower-* or *upper-saddle-node* bifurcation. For $a = -\frac{1}{2}B_1$ with $\Delta_1 = B_1^2 - 4C_1 = 0$, three functional equilibriums are repeated with three multiplicity. The intersected point of $a = -\frac{1}{2}B_1$ with Eq. (3.53) gives

$$a = -\frac{1}{2}(a + \frac{B}{A}). \tag{3.59}$$

Thus,

$$a = -\frac{B}{3A}. \tag{3.60}$$

Such a functional bifurcation at the intersection point is also a third-order *source* or *sink functional* bifurcation. The functional bifurcation diagrams for six cases of three equilibriums with one intersection are presented in Figs. 3.3a–f.

In the 1-dimensional cubic nonlinear functional dynamical system in Eq. (3.1), $(g(x))^2 + B_1 g(x) + C_1 = 0$ for $\Delta_1 = B_1^2 - 4C_1 > 0$ gives two functional equilibriums of $g(x^*) = b_1, b_2$, which have an intersection with $g(x^*) = a$. The intersected point are at $a = b_1$ or $a = b_2$ with Eq. (3.56). The double-repeated functional equilibrium requires $\Delta_1 = B_1^2 - 4C_1 = 0$ and there are two functional equilibriums of $g(x^*) = a, b_1$ under $\Delta_1 = B_1^2 - 4C_1 > 0$ and $x^* = b_2$ for $\Delta_1 = B_1^2 - 4C_1 < 0$. Similarly, there are two functional equilibriums of $g(x^*) = a, b_2$ under $\Delta_1 = B_1^2 - 4C_1 > 0$ and $g(x^*) = b_2$ for $\Delta_1 = B_1^2 - 4C_1 < 0$. Such a bifurcation for two equilibriums appearance and vanishing is called a functional *lower-* or *upper-saddle*-node bifurcation. The stable and unstable equilibriums varying with the vector parameter are also represented by solid and dashed curves, respectively. The functional bifurcation diagrams for four cases of three equilibriums are presented in Figs. 3.4a–d for $a_0 > 0$ and $a_0 < 0$ with $dg/dx|_{x^*} > 0$. If the double-repeated functional equilibrium has an intersection with $g(x^*) = a(\mathbf{p}_0) = -\frac{1}{2}B_1 = -\frac{B}{3A}$, there are two triple-repeated functional equilibriums at $g(x^*) = a(\mathbf{p}_0)$ which are the third-order functional sink and source bifurcations for $a_0 > 0$ and $a_0 < 0$ with $dg/dx|_{x^*} > 0$, respectively. The stability and bifurcation diagrams of functional equilibriums are shown in Figs. 3.4e,f.

Consider a 1-dimensional, cubic nonlinear functional dynamical system with a double-repeated functional equilibrium and one simple functional equilibrium.

(i) For $b < a$, the 1-dimensional, cubic nonlinear functional dynamical system is

$$\dot{x} = a_0(\mathbf{p})(g(x) - b(\mathbf{p}))^2(g(x) - a(\mathbf{p})). \tag{3.61}$$

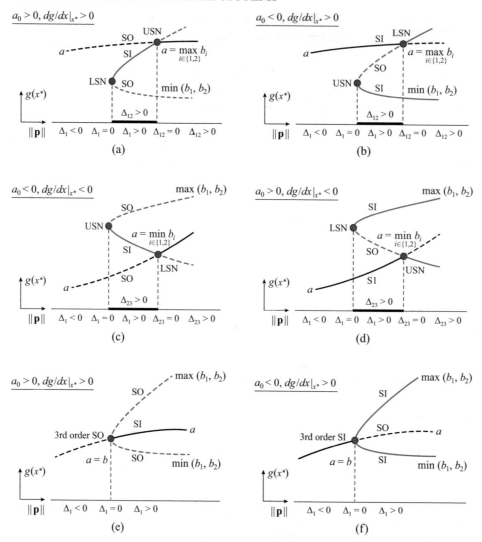

Figure 3.3: Stability and bifurcation of functional equilibriums in the 1-dimensional, cubic nonlinear functional dynamical system: (a) the functional LSN ($\Delta_1 = 0$) and USN ($a = \max\{b_1, b_2\}$) bifurcations ($a_0 > 0, dg/dx|_{x^*} > 0$), (b) the USN ($\Delta_1 = 0$) and LSN ($a = \max\{b_1, b_2\}$) bifurcations ($a_0 < 0, dg/dx|_{x^*} > 0$), (c) the functional USN ($\Delta_1 = 0$) and LSN ($a = \min\{b_1, b_2\}$) bifurcations ($a_0 < 0, dg/dx|_{x^*} < 0$), (d) the functional LSN ($\Delta_1 = 0$) and USN ($a = \min\{b_1, b_2\}$) bifurcations ($a_0 > 0, dg/dx|_{x^*} < 0$), (e) the third-order functional SO bifurcation ($\Delta_1 = 0$ and $a = b$) ($a_0 > 0, dg/dx|_{x^*} > 0$), (f) the third-order functional SI bifurcation ($\Delta_1 = 0$ and $a = b$) ($a_0 < 0, dg/dx|_{x^*} > 0$). LSN: *lower-saddle-node*, USN: *upper-saddle-node*, SI: sink, SO: source. Stable and unstable equilibriums are represented by solid and dashed curves, respectively. The bifurcation points are marked by circular symbols.

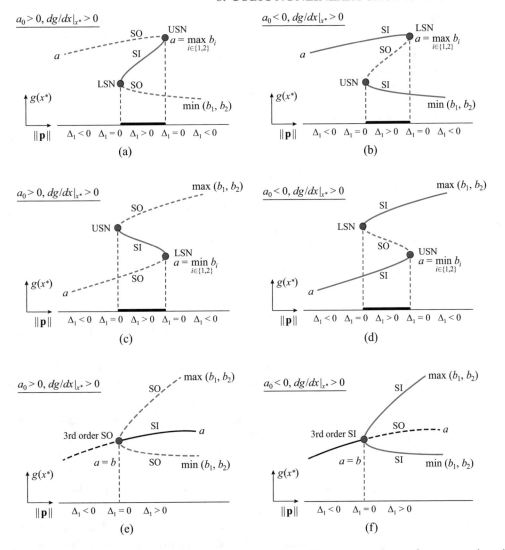

Figure 3.4: Stability and bifurcation of functional equilibriums in the 1-dimensional, cubic nonlinear functional dynamical system: (a) the functional LSN ($\Delta_1 = 0$) and USN ($a = \max\{b_1, b_2\}$) bifurcations ($a_0 > 0$, $dg/dx|_{x^*} > 0$), (b) the functional USN ($\Delta_1 = 0$) and LSN ($a = \max\{b_1, b_2\}$) bifurcations ($a_0 < 0$, $dg/dx|_{x^*} > 0$). (c) the functional USN ($\Delta_1 = 0$) and LSN ($a = \min\{b_1, b_2\}$) bifurcations ($a_0 > 0$, $dg/dx|_{x^*} > 0$), (d) the functional LSN ($\Delta_1 = 0$) and USN ($a = \min\{b_1, b_2\}$) bifurcations ($a_0 < 0$, $dg/dx|_{x^*} > 0$), (e) the third-order functional SO bifurcation ($\Delta_1 = 0$ and $a = b$) ($a_0 > 0$, $dg/dx|_{x^*} > 0$), (f) the third-order functional SI bifurcation ($\Delta_1 = 0$ and $a = b$) ($a_0 < 0$, $dg/dx|_{x^*} > 0$). LSN: lower-saddle-node, USN: *upper-saddle-node*, SI: sink, SO: source. Stable and unstable equilibriums are represented by solid and dashed curves, respectively. The bifurcation points are marked by circular symbols.

For such a system, if $a > 0$ and $dg/dx|_{x*} > 0$, the double-repeated functional equilibrium of $g(x^*) = b$ is a *lower-saddle*, which is unstable, and the simple functional equilibrium of $g(x^*) = b$ is a source, which is unstable. If $a < 0$ and $dg/dx|_{x*} < 0$, the double-repeated functional equilibrium of $g(x^*) = b$ is an *upper-saddle*, which is unstable, and the simple functional equilibrium of $g(x^*) = a$ is a sink, which is stable.

(ii) For $b > a$, the 1-dimensional cubic nonlinear functional dynamical system is

$$\dot{x} = a_0(\mathbf{p})(g(x) - a(\mathbf{p}))(g(x) - b(\mathbf{p}))^2. \tag{3.62}$$

For such a system, if $a_0 > 0$ and $dg/dx|_{x*} > 0$, the double-repeated functional equilibrium of $g(x^*) = b$ is an *upper-saddle*, which is unstable, and the simple functional equilibrium of $g(x^*) = a$ is a source, which is unstable. If $a < 0$ and $dg/dx|_{x*} > 0$, the double-repeated functional equilibrium of $g(x^*) = b$ is a *lower-saddle*, which is unstable, and the simple functional equilibrium of $g(x^*) = a$ is a sink, which is stable.

(iii) For $b = a$, the dynamical system on the boundary is

$$\dot{x} = a_0(\mathbf{p})(g(x) - b(\mathbf{p}))^3. \tag{3.63}$$

For such a system, if $a_0 > 0$ and $dg/dx|_{x*} > 0$, the triple-repeated functional equilibrium of $g(x^*) = b$ with the third-order multiplicity is a functional source bifurcation of the third-order for the (US:SO) to (SO:LS) equilibrium. If $a_0 < 0$ and $dg/dx|_{x*} > 0$, the triple-repeated functional equilibrium of $x^* = b$ with the third-order multiplicity is a functional sink bifurcation of the third-order for the functional (LS:SI) to functional (SO:US) equilibrium. With parameter changes, the bifurcation diagram for the cubic nonlinear system is presented in Fig. 3.5. The acronyms LS, US, SI, and SO are for *lower-saddle*, *upper-saddle*, sink, and source, respectively. Stable and unstable equilibriums are represented by solid and dashed curves, respectively. The bifurcation point is marked by a circular symbol. The third-order functional source bifurcation for the *upper-saddle* and source functional equilibriums to the source and *lower-saddle* functional equilibriums is presented in Fig. 3.5a. The third-order sink bifurcation for the *lower-saddle* and sink equilibriums to the sink and *upper-saddle* equilibriums is presented in Fig. 3.5b.

To illustrate the stability and bifurcation of equilibrium with singularity in a 1-dimensional, cubic nonlinear functional dynamical system, the functional equilibrium of $\dot{x} = a_0(g(x) - a_1)^3$ is presented in Fig. 3.6. The third-order functional sink and source of equilibrium of $x^* = a_1^{(i)} \in S_1$ with $g(x^*) = a_1$ with the third-order multiplicity are stable and unstable for $a_0 dg/dx|_{x*} < 0$ and $a_0 dg/dx|_{x*} > 0$, respectively. The stable and unstable functional equilibriums are depicted by solid and dashed curves for $a_0 dg/dx|_{x*} < 0$ and $a_0 dg/dx|_{x*} > 0$, respectively. At $a_0 = 0$, the third-order functional sink and source equilibriums are switched, which is marked by a circular symbol.

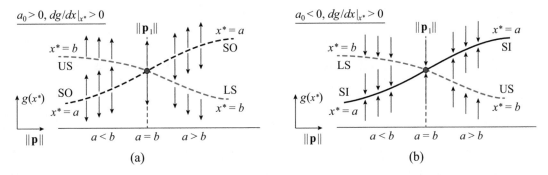

Figure 3.5: Stability and bifurcation of a double-repeated functional equilibrium with a simple functional equilibrium in a 1-dimensional, cubic nonlinear dynamical system: (a) a third-order source bifurcation for (US:SO) to (SO:LS) switching ($a_0 > 0, dg/dx|_{x^*} > 0$), (b) a third-order functional sink bifurcation ($a_0 < 0, dg/dx|_{x^*} > 0$) for (LS:SI) to (SI:US) switching. Stable and unstable equilibriums are represented by solid and dashed curves, respectively.

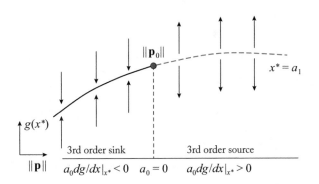

Figure 3.6: Stability of a triple-repeated functional equilibrium in the 1-dimensional, cubic non-linear functional dynamical system: stable and unstable equilibriums are represented by solid and dashed curves, respectively. The stability switching is labeled by a circular symbol.

CHAPTER 4

Quartic Nonlinear Functional Systems

In this chapter, the stability and bifurcation of the quartic nonlinear functional dynamical systems will be presented. The fourth-order functional *upper-saddle* and *lower-saddle* appearing bifurcations of two second-order functional *upper-saddles* and the functional *lower-saddles* will be presented. The third-order sink and source switching bifurcations of *lower-saddle* with sink and *upper-saddle* with source will be discussed.

4.1 QUARTIC FUNCTIONAL SYSTEMS

Definition 4.1 Consider a 1-dimensional, quartic nonlinear functional dynamical system as

$$
\dot{x} = A(\mathbf{p})(g(x))^4 + B(\mathbf{p})(g(x))^3 + C(\mathbf{p})(g(x))^2 + D(\mathbf{p})g(x) + E(\mathbf{p})
$$
$$
= a_0[(g(x))^2 + B_1 g(x) + C_1][(g(x))^2 + B_2 g(x) + C_2], \tag{4.1}
$$

where $A(\mathbf{p}) \neq 0$, and

$$
\mathbf{p} = (p_1, p_2, \ldots, p_m)^{\mathrm{T}}. \tag{4.2}
$$

(i) If

$$
\Delta_i = B_i^2 - 4C_i < 0 \text{ for } i = 1, 2 \tag{4.3}
$$

the quartic nonlinear functional dynamical system does not have any functional equilibrium, and the corresponding standard functional form is

$$
\dot{x} = a_0[(g(x) + \frac{1}{2}B_1)^2 + \frac{1}{4}(-\Delta_1)][(g(x) + \frac{1}{2}B_2)^2 + \frac{1}{4}(-\Delta_2)]. \tag{4.4}
$$

The flow of such a functional system without functional equilibriums is called a non-equilibrium flow.

(i_1) If $a_0 > 0$, the non-equilibrium flow is called the positive flow.

(i_2) If $a_0 < 0$, the non-equilibrium flow is called the negative flow.

(ii) If

$$
\Delta_i = B_i^2 - 4C_i > 0 \text{ and } \Delta_j = B_j^2 - 4C_j < 0 \text{ for } i, j \in \{1, 2\}, \ i \neq j \tag{4.5}
$$

the quartic nonlinear functional system has two simple functional equilibrium sets, i.e.,

$$x^* \in S_1 = \{a_1^{(s)} | g(a_1^{(s)}) = a_1, \ s = 1, 2, \ldots, N_1\},$$
$$x^* \in S_2 = \{a_2^{(s)} | g(a_2^{(s)}) = a_2, \ s = 1, 2, \ldots, N_2\},$$
$$\text{with } \{a_1, a_2\} = \text{sort } \{b_1^{(i)}, b_2^{(i)}\}, \ a_1 < a_2; \tag{4.6}$$
$$b_1^{(i)} = -\frac{1}{2}(B_i + \sqrt{\Delta_i}), \ b_2^{(i)} = -\frac{1}{2}(B_i - \sqrt{\Delta_i}).$$

The corresponding standard functional form is

$$\dot{x} = a_0(g(x) - a_1)(g(x) - a_2)[(g(x) + \frac{1}{2}B_j)^2 + \frac{1}{4}(-\Delta_j)], \tag{4.7}$$

where

$$a_1 = \min\{b_1^{(i)}, b_2^{(i)}\} \text{ and } a_2 = \max\{b_1^{(i)}, b_2^{(i)}\}. \tag{4.8}$$

Such a flow of equilibriums is called a flow of two simple functional equilibriums.

(iii) If

$$\Delta_i = B_i^2 - 4C_i = 0 \text{ and } \Delta_j = B_j^2 - 4C_j < 0 \text{ for } i, j \in \{1, 2\}, \ i \neq j \tag{4.9}$$

the quartic nonlinear functional dynamical system has a double-repeated functional equilibrium set, i.e.,

$$x^* \in S_1 = \{a_1^{(s)} | g(a_1^{(s)}) = a_1, \ s = 1, 2, \ldots, N_1\},$$
$$a_1 = b_1^{(i)} = b_2^{(i)} = -\frac{1}{2}B_i. \tag{4.10}$$

The corresponding standard functional form is

$$\dot{x} = a_0(g(x) - a_1)^2[(g(x) + \frac{1}{2}B_j)^2 + \frac{1}{4}(-\Delta_j)], \tag{4.11}$$

where

$$a_1 = b_1^{(i)} = b_2^{(i)}. \tag{4.12}$$

Such a flow of the equilibrium of $x^* = a_1^{(i)} \in S_1$ with $g(x^*) = a_1$ is called a *saddle* flow of the second order. The functional equilibrium of $g(x^*) = a_1$ for two functional equilibriums switching or appearance or vanishing is called a bifurcation point of equilibrium at $\mathbf{p} = \mathbf{p}_1 \in \partial\Omega_{12}$, and the functional bifurcation condition is

$$\Delta_i = B_i^2 - 4C_i = 0 \ (i \in \{1, 2\}) \text{ and } a_1 = -\frac{1}{2}B_i. \tag{4.13}$$

(iv) If

$$\Delta_i = B_i^2 - 4C_i \geq 0 \text{ for } i = 1, 2. \tag{4.14}$$

The quartic nonlinear functional dynamical system has four functional equilibriums, i.e.,

$$x^* \in S_\alpha = \{a_\alpha^{(s)} | g(a_1^{(s)}) = a_\alpha, s = 1, 2, \ldots, N_\alpha\},$$

$$\text{with } \{\cup_{\alpha=1}^4 a_\alpha\} = \text{sort}\{\cup_{i=1}^2 \{b_1^{(i)}, b_2^{(i)}\}\}, a_\alpha < a_{\alpha+1}; \tag{4.15}$$

$$b_1^{(i)} = -\frac{1}{2}(B_i + \sqrt{\Delta_i}), \; b_2^{(i)} = -\frac{1}{2}(B_i - \sqrt{\Delta_i}), \text{ for } i = 1, 2.$$

(iv$_1$) A standard functional form is

$$\dot{x} = a_0 \prod_{i=1}^4 (g(x) - a_i), \tag{4.16}$$

where

$$\Delta_i = B_i^2 - 4C_i > 0, \; i = 1, 2;$$

$$b_k^{(1)} \neq b_l^{(2)} \text{ for } k, l \in \{1, 2\}; \tag{4.17}$$

$$a_{1,2,3,4} \in \{b_1^{(1)}, b_2^{(1)}, b_1^{(2)}, b_2^{(2)}\} \text{ with } a_m < a_{m+1}.$$

Such a flow of functional equilibriums is called a flow of four simple functional equilibriums.

(iv$_2$) The corresponding standard functional form is

$$\dot{x} = a_0(g(x) - a_{i_1})^2(g(x) - a_{i_2})(g(x) - a_{i_3}), \tag{4.18}$$

where

$$\Delta_i = B_i^2 - 4C_i > 0, \; \Delta_j = B_j^2 - 4C_j > 0 \text{ for } i, j = 1, 2;$$

$$a_{i_1} = b_k^{(i)} = b_l^{(j)}, \; (i, k) \neq (j, l); \; i, j, k, l \in \{1, 2\}; \tag{4.19}$$

$$a_{i_1} \notin \{a_{i_2}, a_{i_3}, \} \text{ for } i_\alpha \in \{1, 2, 3, 4\} \text{ and } \alpha \in \{1, 2, 3, 4\}.$$

Such a flow of equilibrium of $x^* = a_{i_1}^{(s)} \in S_{i_1}$ with $g(x^*) = a_{i_1}$ is called a functional *saddle* flow of the second order. The functional equilibrium of $g(x^*) = a_{i_1}$ for two functional equilibriums switching or appearance or vanishing is called a functional *saddle* bifurcation of equilibrium at a point $\mathbf{p} = \mathbf{p}_1 \in \partial\Omega_{12}$, and the functional bifurcation condition is

$$\Delta_i = B_i^2 - 4C_i > 0 \, (i \in \{1, 2\}) \text{ and}$$

$$\Delta_j = B_j^2 - 4C_j > 0 \, (j \in \{1, 2\}), \tag{4.20}$$

$$b_k^{(i)} = b_l^{(j)}, \; (i, k) \neq (j, l), \; (i, j, k, l \in \{1, 2\}).$$

(iv₃) The corresponding standard functional form is

$$\dot{x} = a_0(g(x) - a_{i_1})^3(g(x) - a_{i_2}), \tag{4.21}$$

where

$$\Delta_i = B_i^2 - 4C_i > 0, \ \Delta_j = B_j^2 - 4C_j = 0 \text{ for } i, j = 1, 2;$$

$$a_{i_1} = -\frac{1}{2}B_j = b_l^{(i)}, \ a_{i_2} = b_k^{(i)}, \ k \neq l; k, l \in \{1, 2\}; \tag{4.22}$$

$$a_{i_1} = a_{i_3} \text{ for } i_\alpha \in \{1, 2, 3, 4\} \text{ and } \alpha \in \{1, 2, 3, 4\}.$$

Such a flow of the functional equilibrium of $x^* = a_{i_1}^{(s)} \in S_{i_1}$ with $g(x^*) = a_{i_1}$ is called a functional source or sink flow of the third order. The functional equilibrium of $x^* = a_{i_1}^{(s)} \in S_{i_1}$ with $g(x^*) = a_{i_1}$ for one functional equilibrium to three functional equilibrium is called a third-order source or sink bifurcation of functional equilibrium at a point $\mathbf{p} = \mathbf{p}_1 \in \partial\Omega_{12}$, and the bifurcation condition is

$$\Delta_i = B_i^2 - 4C_i > 0 \, (i \in \{1, 2\}) \text{ and}$$
$$\Delta_j = B_j^2 - 4C_j > 0 \, (j \in \{1, 2\}); \tag{4.23}$$
$$b_k^{(i)} = -\frac{1}{2}B_j, \ b_k^{(i)}, \neq b_l^{(i)}, \ (k \neq l, \ k, l \in \{1, 2\}).$$

(iv₄) The corresponding standard functional form is

$$\dot{x} = a_0(g(x) - a_1)^2(g(x) - a_2)^2, \tag{4.24}$$

where

$$\Delta_i = B_i^2 - 4C_i = 0, \ i = 1, 2$$

$$b_1^{(1)} = b_2^{(1)} = -\frac{1}{2}B_1, \ b_1^{(2)} = b_2^{(2)} = -\frac{1}{2}B_2, \ B_1 \neq B_2; \tag{4.25}$$

$$a_1 = \min\{-\frac{1}{2}B_1, -\frac{1}{2}B_2\}, \ a_2 = \max\{-\frac{1}{2}B_1, -\frac{1}{2}B_2\}.$$

Such a flow with the two functional equilibriums of $x^* = a_1^{(s)} \in S_1$ with $g(x^*) = a_1$ and $x^* = a_2^{(s)} \in S_2$ with $g(x^*) = a_2$ is called a (LS:LS)- or (US:US)-flow. The functional equilibriums of $x^* = a_1^{(s)} \in S_1$ with $g(x^*) = a_1$ and $x^* = a_2^{(s)} \in S_2$ with $g(x^*) = a_2$ for two functional equilibriums switching are called two bifurcations of functional equilibrium at a point $\mathbf{p} = \mathbf{p}_1 \in \partial\Omega_{12}$, and the bifurcation condition is

$$\Delta_i = B_i^2 - 4C_i = 0, \ i = 1, 2;$$
$$b_1^{(1)} = b_2^{(1)} = -\frac{1}{2}B_1, \ b_1^{(2)} = b_2^{(2)} = -\frac{1}{2}B_2. \tag{4.26}$$

(iv$_5$) The corresponding standard functional form is

$$\dot{x} = a_0(g(x) - a_1)^4, \tag{4.27}$$

where

$$\Delta_i = B_i^2 - 4C_i = 0, \ i = 1, 2;$$
$$b_1^{(i)} = b_2^{(i)} = -\frac{1}{2}B_i, \ B_1 = B_2. \tag{4.28}$$

Such a flow at the functional equilibrium of $x^* = a_1^{(s)} \in S_1$ with $g(x^*) = a_1$ is called a functional *saddle* flow of the fourth order. The functional equilibrium of $x^* = a_1^{(s)} \in S_1$ with $g(x^*) = a_1$ for two second-order functional equilibriums switching or four equilibrium appearance is called a fourth-order *saddle* bifurcation of functional equilibrium at a point $\mathbf{p} = \mathbf{p}_1 \in \partial\Omega_{12}$, and the bifurcation condition is

$$\Delta_i = B_i^2 - 4C_i = 0, \ a_1 = b_1^{(i)} = b_2^{(i)}, \ i = 1, 2. \tag{4.29}$$

Theorem 4.2

(i) *Under conditions of*

$$\Delta_i = B_i^2 - 4C_i < 0, \ for \ i = 1, 2, \tag{4.30}$$

a standard functional form of Eq. (4.1) *is*

$$\dot{x} = f(x, \mathbf{p}) = a_0[(g(x) + \frac{1}{2}B_1)^2 + \frac{1}{4}(-\Delta_1)][(g(x) + \frac{1}{2}B_2)^2 + \frac{1}{4}(-\Delta_2)] \tag{4.31}$$

with $a_0 = A(\mathbf{p})$, *which has a non-functional equilibrium flow.*

(i$_1$) *If* $a_0(\mathbf{p}) > 0$, *the non-functional equilibrium flow is called a positive flow.*

(i$_2$) *If* $a_0(\mathbf{p}) > 0$, *the non-functional equilibrium flow is called a negative flow.*

(ii) *Under a condition of*

$$\Delta_i = B_i^2 - 4C_i > 0 \ and \ \Delta_j = B_j^2 - 4C_j < 0 \ for \ i, j \in \{1, 2\}, \ i \neq j, \tag{4.32}$$

a standard functional form of Eq. (4.1) *is*

$$\dot{x} = f(x, \mathbf{p}) = a_0(g(x) - a_1)(g(x) - a_2)[(g(x) + \frac{1}{2}B_j)^2 + \frac{1}{4}(-\Delta_j)], \tag{4.33}$$

where

$$a_1 = \min\{b_1^{(i)}, b_2^{(i)}\} \ and \ a_2 = \max\{b_1^{(i)}, b_2^{(i)}\},$$
$$b_1^{(i)} = -\frac{1}{2}(B_i + \sqrt{\Delta_i}), \ b_2^{(i)} = -\frac{1}{2}(B_i - \sqrt{\Delta_i}). \tag{4.34}$$

(ii₁) *The functional equilibriums of $x^* = a_1^{(s)} \in S_1$ with $g(x^*) = a_1$ and $x^* = a_2^{(s)} \in S_2$ with $g(x^*) = a_2$ are stable (sink, $df/dx|_{x^*=a_1} < 0$) and unstable (source, $df/dx|_{x^*=a_2} > 0$), respectively.*

(ii₂) *The functional equilibriums of $x^* = a_1^{(s)} \in S_1$ with $g(x^*) = a_1$ and $x^* = a_2^{(s)} \in S_2$ with $g(x^*) = a_2$ are unstable (source, $df/dx|_{x^*=a_1} > 0$) and stable (sink, $df/dx|_{x^*=a_2} < 0$), respectively.*

(iii) *Under conditions of*

$$\Delta_i = B_i^2 - 4C_i = 0 \text{ and } \Delta_j = B_j^2 - 4C_j < 0 \text{ for } i, j \in \{1, 2\}, \ i \neq j, \quad (4.35)$$

a standard functional form of Eq. (4.1) is

$$\dot{x} = f(x, \mathbf{p}) = a_0 (g(x) - a_1)^2 [(g(x) + \tfrac{1}{2}B_j)^2 + \tfrac{1}{4}(-\Delta_j)], \quad (4.36)$$

where

$$a_1 = b_1^{(i)} = b_2^{(i)} = -\frac{1}{2}B_i. \quad (4.37)$$

(iii₁) *The functional equilibrium of $x^* = a_1^{(s)} \in S_1$ with $g(x^*) = a_1$ is unstable (a functional upper-saddle, $d^2 f/dx^2|_{x^*=a_1} > 0$ and $dg/dx|_{x^*} \neq 0$). Such a flow at the functional equilibrium of $x^* = a_1^{(s)} \in S_1$ with $g(x^*) = a_1$ is called a functional upper-saddle flow of the second order. The bifurcation of functional equilibrium of $x^* = a_1^{(s)} \in S_1$ with $g(x^*) = a_1$ for two functional equilibriums appearance or vanishing is called a functional upper-saddle-node bifurcation of the second order at a point $\mathbf{p} = \mathbf{p}_1 \in \partial\Omega_{12}$.*

(iii₂) *The functional equilibrium of $x^* = a_1^{(s)} \in S_1$ with $g(x^*) = a_1$ is unstable (a functional lower-saddle, $d^2 f/dx^2|_{x^*=a_1} < 0$ and $dg/dx|_{x^*} \neq 0$). Such a flow at the functional equilibrium of $x^* = a_1^{(s)} \in S_1$ with $g(x^*) = a_1$ is called a functional lower-saddle flow of the second order. The bifurcation of functional equilibrium of $x^* = a_1^{(s)} \in S_1$ with $g(x^*) = a_1$ for two functional equilibriums appearance or vanishing is called a functional lower-saddle-node bifurcation of the second order at a point $\mathbf{p} = \mathbf{p}_1 \in \partial\Omega_{12}$.*

(iv) *Under conditions of*

$$\begin{aligned}
&\Delta_i = B_i^2 - 4C_i > 0, \ i = 1, 2; \\
&b_k^{(1)} \neq b_l^{(2)} \text{ for } k, l \in \{1, 2\}; \\
&b_1^{(i)} = -\frac{1}{2}(B_i + \sqrt{\Delta_i}), \ b_2^{(i)} = -\frac{1}{2}(B_i - \sqrt{\Delta_i}), \text{ for } i = 1, 2,
\end{aligned} \quad (4.38)$$

a standard functional form is

$$\dot{x} = f(x, \mathbf{p}) = a_0(g(x) - a_1)(g(x) - a_2)(g(x) - a_3)(g(x) - a_4), \qquad (4.39)$$

where

$$a_{1,2,3,4} \in \cup_{i=1}^{2}\{b_1^{(i)}, b_2^{(i)}\} \text{ with } a_m < a_{m+1}. \qquad (4.40)$$

(iv_1) For $a_0(\mathbf{p})dg/dx|_{x^*} > 0$, the simple functional equilibriums of $x^* = a_i^{(s)} \in S_i$ with $g(x^*) = a_i$ ($i = 1, 2, 3, 4$) are stable, unstable, stable, and unstable, respectively. The flow is called a (SI:SO:SI:SO) flow.

(iv_2) For $a_0(\mathbf{p})dg/dx|_{x^*} < 0$, the simple functional equilibriums of $x^* = a_i^{(s)} \in S_i$ with $g(x^*) = a_i$ ($i = 1, 2, 3, 4$) are unstable, stable, unstable, and stable, respectively. The flow is called a (SO:SI:SO:SI) flow.

The simple functional equilibrium of $x^* = a_i^{(s)} \in S_i$ with $g(x^*) = a_i$ ($i = 1, 2, 3, 4$) is unstable (source, $df/dx|_{x^*} > 0$) and stable (sink, $df/dx|_{x^*} < 0$).

(v) Under conditions of

$$\Delta_i = B_i^2 - 4C_i > 0 \ (i \in \{1, 2\}) \text{ and}$$
$$\Delta_j = B_j^2 - 4C_j > 0 \ (j \in \{1, 2\}),$$
$$b_1^{(\alpha)} = -\frac{1}{2}(B_\alpha + \sqrt{\Delta_\alpha}), \ b_2^{(\alpha)} = -\frac{1}{2}(B_\alpha - \sqrt{\Delta_\alpha}), \text{ for } \alpha = i, j; \qquad (4.41)$$
$$b_k^{(i)} = b_l^{(j)}, \ (i, k) \neq (j, l), \ (i, j, k, l \in \{1, 2\}),$$

a standard functional form of Eq. (4.1) is

$$\dot{x} = f(x, \mathbf{p}) = a_0(g(x) - a_{i_1})^2(g(x) - a_{i_2})(g(x) - a_{i_3}), \qquad (4.42)$$

where

$$a_{i_1} = b_k^{(i)} = b_l^{(j)} \in \cup_{i=1}^{2}\{b_1^{(i)}, b_1^{(i)}\}, \ (i, k) \neq (j, l); \ i, j, k, l \in \{1, 2\}, \qquad (4.43)$$
$$a_{i_1} \notin \{a_{i_2}, a_{i_3}\} \subset \cup_{i=1}^{2}\{b_1^{(i)}, b_1^{(i)}\}, \text{ for } i_\alpha \in \{1, 2, 3\} \text{ and } \alpha \in \{1, 2, 3\}.$$

(v_1) The functional equilibriums of $x^* = a_\alpha^{(s)} \in S_\alpha$ with $g(x^*) = a_\alpha(\alpha = i_2, i_3)$ are unstable (source, $df/dx|_{x^*} > 0$) and stable (sink, $df/dx|_{x^*} < 0$).

(v_2) The functional equilibrium of $x^* = a_{i_1}^{(s)} \in S_{i_1}$ with $g(x^*) = a_{i_1}$ is unstable (a functional upper-saddle, $d^2f/dx^2|_{x^*} > 0$ and $dg/dx|_{x^*} \neq 0$) and unstable (a functional lower-saddle, $d^2f/dx^2|_{x^*} < 0$ and $dg/dx|_{x^*} \neq 0$). The bifurcation of functional equilibrium at $x^* = a_{i_1}^{(s)} \in S_{i_1}$ with $g(x^*) = a_{i_1}$ for two functional equilibriums switching or vanishing is called the functional upper-saddle-node or lower-saddle-node bifurcation of the second order at a point $\mathbf{p} = \mathbf{p}_1 \in \partial\Omega_{12}$.

(vi) Under conditions of

$$\Delta_i = B_i^2 - 4C_i > 0 \, (i \in \{1, 2\}) \text{ and}$$
$$\Delta_j = B_j^2 - 4C_j = 0 \, (j \in \{1, 2\});$$
$$b_1^{(i)} = -\frac{1}{2}(B_i + \sqrt{\Delta_i}), \; b_2^{(i)} = -\frac{1}{2}(B_i + \sqrt{\Delta_i}), \qquad (4.44)$$
$$b_{1,2}^{(j)} = -\frac{1}{2}B_j, \; b_k^{(i)} = b_l^{(i)}, \; (k \neq l, \, k, l \in \{1, 2\}),$$

a standard functional form of Eq. (4.1) *is*

$$\dot{x} = f(x, \mathbf{p}) = a_0(g(x) - a_{i_1})^3(g(x) - a_{i_2}), \qquad (4.45)$$

where

$$a_{i_1} = b_{1,2}^{(j)} = b_l^{(i)}, \; a_{i_2} = b_k^{(i)}, \; a_1 < a_2;$$
$$\text{for } i, j, l \in \{1, 2\}, \; i_\alpha \in \{1, 2\} \text{ and } \alpha \in \{1, 2\}. \qquad (4.46)$$

(vi$_1$) The functional equilibriums of $x^* = a_{i_2}^{(s)} \in S_{i_2}$ with $g(x^*) = a_{i_2}$ are unstable (source, $df/dx|_{x^*} > 0$) and stable (sink, $df/dx|_{x^*} < 0$).

(vi$_2$) The functional equilibrium of $x^* = a_{i_2}^{(s)} \in S_{i_1}$ with $g(x^*) = a_{i_1}$ is unstable (the third-order source, $d^3 f/dx^3|_{x^*} > 0$ and $dg/dx|_{x^*} \neq 0$) and stable (the third-order sink, $d^3 f/dx^3|_{x^*} < 0$ and $dg/dx|_{x^*} \neq 0$). The bifurcation of functional equilibrium at $x^* = a_{i_1}^{(s)} \in S_{i_1}$ with $g(x^*) = a_{i_1}$ for one functional equilibrium to three functional equilibrium is called the functional source or sink bifurcation of the third order at a point $\mathbf{p} = \mathbf{p}_1 \in \partial\Omega_{12}$.

(vii) Under conditions of

$$\Delta_i = B_i^2 - 4C_i = 0 \, (i \in \{1, 2\}) \text{ and}$$
$$\Delta_j = B_j^2 - 4C_j = 0 \, (j \in \{1, 2\}),$$
$$b_1^{(\alpha)} = b_2^{(\alpha)} = -\frac{1}{2}B_\alpha \text{ for } \alpha = i, j; \qquad (4.47)$$
$$B_1 \neq B_2,$$

a standard functional form of Eq. (4.1) *is*

$$\dot{x} = a_0(g(x) - a_1)^2(g(x) - a_2)^2, \qquad (4.48)$$

where

$$a_1 = \min\{-\frac{1}{2}B_1, -\frac{1}{2}B_2\}, \; a_2 = \max\{-\frac{1}{2}B_1, -\frac{1}{2}B_2\}. \qquad (4.49)$$

(vii_1) *For $a_0(\mathbf{p}) > 0$, the functional equilibriums of $x^* = a_i^{(s)} \in S_i$ with $g(x^*) = a_i$ ($i = 1, 2$) are unstable (upper-saddle, $d^2 f/dx^2|_{x*} > 0$ and $dg/dx|_{x*} \neq 0$). The functional equilibrium of $x^* = a_i^{(s)} \in S_i$ with $g(x^*) = a_i$ for two functional equilibriums vanishing and appearance are called a functional upper-saddle-node bifurcation of the second order at a point $\mathbf{p} = \mathbf{p}_1 \in \partial\Omega_{12}$.*

(vii_2) *For $a_0(\mathbf{p}) < 0$, the functional equilibriums of $x^* = a_i^{(s)} \in S_i$ with $g(x^*) = a_i$ ($i = 1, 2$) are unstable (lower-saddle, $d^2 f/dx^2|_{x*=a_i} < 0$ and $dg/dx|_{x*} \neq 0$). The functional equilibrium of $x^* = a_i^{(s)} \in S_i$ with $g(x^*) = a_i$ for two functional equilibriums vanishing and appearance are called a functional lower-saddle-node bifurcation of the second order at a point $\mathbf{p} = \mathbf{p}_1 \in \partial\Omega_{12}$.*

($viii$) *Under conditions of*

$$\Delta_i = B_i^2 - 4C_i = 0 \, (i \in \{1, 2\}) \text{ and}$$
$$\Delta_j = B_j^2 - 4C_j = 0 \, (j \in \{1, 2\});$$
$$b_1^{(\alpha)} = b_2^{(\alpha)} = -\frac{1}{2}B_\alpha \text{ for } \alpha = i, j; \tag{4.50}$$
$$B_1 \neq B_2,$$

the corresponding standard functional form is

$$\dot{x} = a_0(g(x) - a_1)^4, \tag{4.51}$$

where

$$a_1 = -\frac{1}{2}B_1 = -\frac{1}{2}B_2. \tag{4.52}$$

($viii_1$) *The functional equilibriums of $x^* = a_1^{(s)} \in S_1$ with $g(x^*) = a_1$ are unstable (functional upper-saddle, $d^4 f/dx^4|_{x*=a_1} > 0$ and $dg/dx|_{x*} \neq 0$). The functional equilibrium of $x^* = a_1^{(s)} \in S_1$ with $g(x^*) = a_1$ for four functional equilibriums vanishing and appearance are called a functional upper-saddle-node bifurcation of the fourth order at a point $\mathbf{p} = \mathbf{p}_1 \in \partial\Omega_{12}$.*

($viii_2$) *The functional equilibriums of $x^* = a_1^{(s)} \in S_1$ with $g(x^*) = a_1$ are unstable (functional lower-saddle, $d^4 f/dx^4|_{x*=a_1} < 0$ and $dg/dx|_{x*} \neq 0$). The equilibrium of $x^* = a_1$ for four functional simple equilibriums vanishing and appearance are called a functional lower-saddle-node bifurcation of the fourth order at a point $\mathbf{p} = \mathbf{p}_1 \in \partial\Omega_{12}$.*

From the previous discussion, a quartic nonlinear functional system is expressed by the product of two quadratic polynomials, i.e.,

$$\dot{x} = a_0[(g(x))^2 + B_1 g(x) + C_1][(g(x))^2 + B_2 g(x) + C_2]. \tag{4.53}$$

Thus, for $\dot{x} = 0$, the functional equilibriums are determined by the roots of two quadratic polynomial equations, i.e.,

$$(g(x))^2 + B_1 g(x) + C_1 = 0 \text{ and/or } (g(x))^2 + B_2 g(x) + C_2 = 0. \tag{4.54}$$

The roots of such quadratic functional polynomial equations are determined by the corresponding discriminant of the quadratic equations, i.e.,

$$\Delta_i = B_i^2 - 4C_i \text{ for } i = 1, 2. \tag{4.55}$$

If $\Delta_i < 0$, the quadratic equation of $(g(x))^2 + B_i g(x) + C_i = 0$ does not have any roots. If $\Delta_i > 0$, the quadratic equation of $(g(x))^2 + B_i g(x) + C_i = 0$ possesses two roots. If $\Delta_i = 0$, the quadratic equation of $(g(x))^2 + B_i g(x) + C_i = 0$ has a repeated root. With parameter variation, suppose one of two quadratic polynomial equations has one root intersected with the roots of the other quadratic polynomial equation. There are six cases for $a_0 > 0$ and $dg/dx|_{x*} > 0$:
(a) $b_2^{(i)} = b_1^{(j)}$, (b) $b_1^{(j)} = b_1^{(i)} = b_2^{(i)} = -\frac{1}{2}B_i$. (c) $b_1^{(i)} = b_1^{(j)}$, (d) $b_2^{(i)} = b_2^{(j)}$, (e) $b_2^{(j)} = b_1^{(i)} = b_2^{(i)} = -\frac{1}{2}B_i$, (f) $b_1^{(i)} = b_2^{(j)}$, as presented in Fig. 4.1. The intersected point for non-repeated roots is a *saddle-node* bifurcation for the subcritical case. The functional *lower-saddle-node* and *upper-saddle-node* bifurcations are shown in Figs. 4.1a,b, and d,f, respectively. The bifurcation dynamics for the 1-dimensional quartic nonlinear functional system is determined by $\dot{x} = a_0(g(x) - a_{i_1})^2(g(x) - a_{i_2})(g(x) - a_{i_3})$ with $i_\alpha, \alpha \in \{1, 2, 3\}$ for four functional equilibriums or $\dot{x} = a_0(g(x) - a_i)^2[(g(x) + \frac{1}{2}B_j)^2 - \frac{1}{4}\Delta_j]$ with $i, j \in \{1, 2\}$ for two functional equilibriums. If the intersected point occurs at the repeated root, the third-order functional source and sink bifurcations are presented in Figs. 4.1b and d, respectively. The corresponding bifurcation dynamics for the 1-dimensional quartic nonlinear functional system is determined by $\dot{x} = a_0(g(x) - a_{i_1})^3(g(x) - a_{i_2})$ with $i_\alpha, \alpha \in \{1, 2\}$. The stable and unstable equilibriums are presented by the solid and dashed curves, respectively. The intersected points are marked by circular symbols, which are the bifurcation points. Without losing generality, suppose the two roots of the quadratic polynomial equation have a relation of $b_1^{(i)} > b_2^{(i)}$ for $i = 1, 2$. The repeated roots of the two functional quadratic polynomial equations are also the *upper-* or *lower-saddle-node* bifurcations for two functional equilibriums appearance and vanishing. Similarly, the six cases of stability and bifurcation diagrams varying with parameter for $a_0 < 0$ and $dg/dx|_{x*} > 0$ are presented in Fig. 4.2. The stability and bifurcation conditions for $a_0 < 0$ and $dg/dx|_{x*} > 0$ are opposite to $a_0 > 0$ and $dg/dx|_{x*} > 0$.

If the roots of two quadratic functional equations do not have any intersections, the open loops for stability and bifurcation diagrams of equilibriums for $a_0 > 0$ and $dg/dx|_{x*} > 0$ with $a_0 < 0$ and $dg/dx|_{x*} > 0$ are presented in Fig. 4.3. There are four cases of open loops for $a_0 > 0, dg/dx|_{x*} > 0$: (a) $B_i < B_j$, (b) $B_i > B_j$, (c) $b_2^{(j)} < -\frac{1}{2}B_i < b_1^{(j)}$, (d) $\Delta_i = \Delta_j, B_i \neq B_j$ and four cases of open loops for $a_0 < 0$ and $dg/dx|_{x*} > 0$: (e) $B_i < B_j$, (f) $B_i > B_j$, (g) $b_2^{(j)} < -\frac{1}{2}B_i < b_1^{(j)}$, (h) $\Delta_i = \Delta_j, B_i \neq B_j$. The two bifurcations occur at the same time because the quadratic functional equations have $\Delta_i = \Delta_j, B_i \neq B_j$. The bifurcation points are only

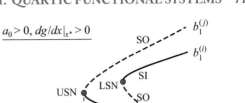

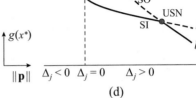

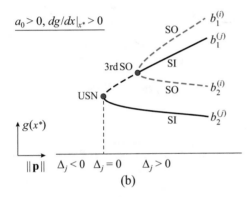

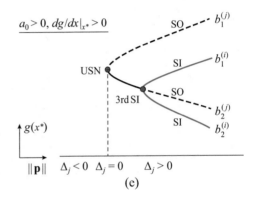

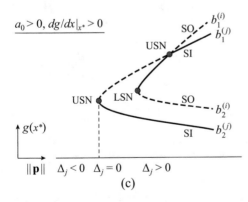

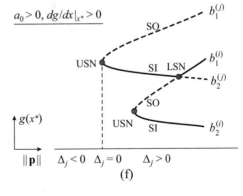

Figure 4.1: Stability and bifurcations of functional equilibriums in the 1-dimensional, quartic nonlinear dynamical system ($a_0 > 0$, $dg/dx|_{x^*} > 0$): (a) $b_2^{(i)} = b_1^{(j)}$, (b) $b_1^{(j)} = b_1^{(i)} = b_2^{(i)} = -\frac{1}{2}B_i$. (c) $b_1^{(i)} = b_1^{(j)}$, (d) $b_2^{(i)} = b_2^{(j)}$, (e) $b_2^{(j)} = b_1^{(i)} = b_2^{(i)} = -\frac{1}{2}B_i$, (f) $b_1^{(i)} = b_2^{(j)}$. LSN: *lower-saddle-node*, USN: *upper-saddle-node*, SI: sink, SO: source. Stable and unstable equilibriums are represented by solid and dashed curves, respectively. The bifurcation points are marked by circular symbols.

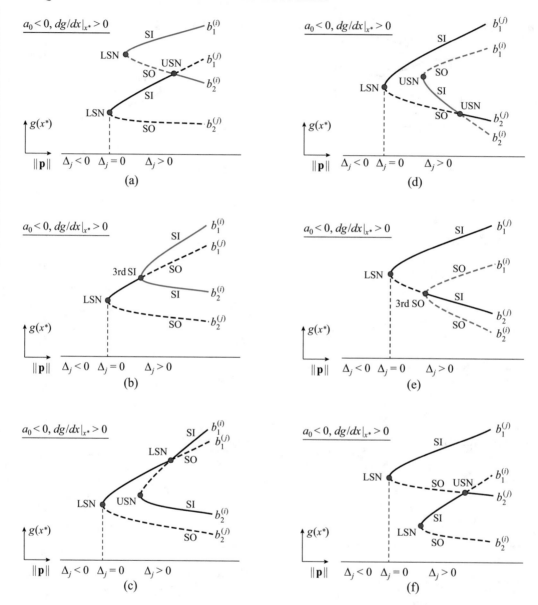

Figure 4.2: Stability and bifurcations of functional equilibriums in the 1-dimensional, quartic nonlinear dynamical system ($a_0 < 0$, $dg/dx|_{x^*} > 0$): (a) $b_2^{(i)} = b_1^{(j)}$, (b) $b_1^{(j)} = b_1^{(i)} = b_2^{(i)} = -\frac{1}{2}B_i$. (c) $b_1^{(i)} = b_1^{(j)}$, (d) $b_2^{(i)} = b_2^{(j)}$, (e) $b_2^{(j)} = b_1^{(i)} = b_2^{(i)} = -\frac{1}{2}B_i$, (f) $b_1^{(i)} = b_2^{(j)}$. LSN: *lower-saddle-node*, USN: *upper-saddle-node*, SI: sink, SO: source. Stable and unstable equilibriums are represented by solid and dashed curves, respectively. The bifurcation points are marked by circular symbols.

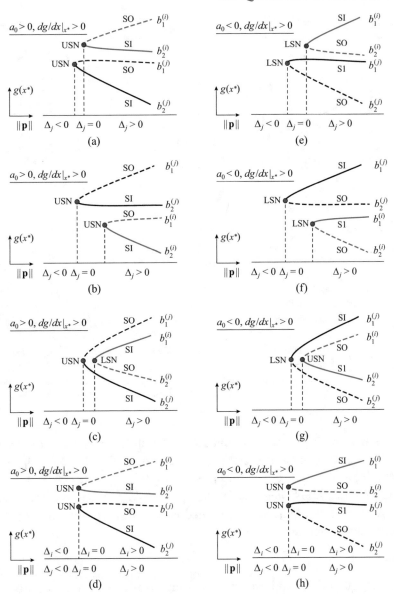

Figure 4.3: Stability and bifurcation of equilibriums in the 1-dimensional, quartic nonlinear dynamical system. $(a_0 > 0, dg/dx|_{x^*} > 0)$: (a) $B_i < B_j$, (b) $B_i > B_j$, (c) $b_2^{(j)} < -\frac{1}{2}B_i < b_1^{(j)}$, (d) $\Delta_i = \Delta_j$, $B_i \neq B_j$. $(a_0 < 0, dg/dx|_{x^*} > 0)$: (e) $B_i < B_j$, (f) $B_i > B_j$, (g) $b_2^{(j)} < -\frac{1}{2}B_i < b_1^{(j)}$, (h) $\Delta_i = \Delta_j$, $B_i \neq B_j$. LSN: *lower-saddle-node*, USN: *upper-saddle-node*, SI-sink, SO-source. Stable and unstable equilibriums are represented by solid and dashed curves, respectively. The bifurcation points are marked by circular symbols.

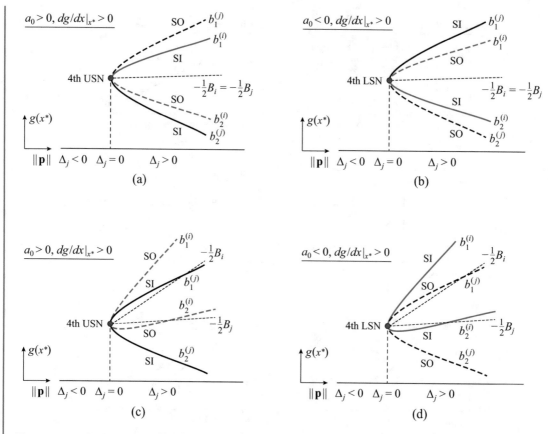

Figure 4.4: Stability and bifurcation of functional equilibriums in the 1-dimensional, quartic nonlinear functional dynamical system. ($\Delta_i > \Delta_j, B_i = B_j$): (a) $a_0 > 0$, $dg/dx|_{x*} > 0$, (b) $a_0 < 0$, $dg/dx|_{x*} > 0$; ($B_i < B_j$): (c) $a_0 > 0$, $dg/dx|_{x*} > 0$, (d) $a_0 < 0$, $dg/dx|_{x*} > 0$. LSN: *lower-saddle-node*, USN: *upper-saddle-node*, SI: sink, SO: source. Stable and unstable equilibriums are represented by solid and dashed curves, respectively. The bifurcation points are marked by circular symbols.

for two functional equilibriums appearance or vanishing from the discriminants of the quadratic equations. The bifurcation dynamics for the 1-dimensional, quartic nonlinear functional dynamical system is determined by $\dot{x} = a_0(g(x) - a_{i_1})^2(g(x) - a_{i_2})(g(x) - a_{i_3})$ with $i_\alpha, \alpha \in \{1, 2, 3\}$.

If $\Delta_i = \Delta_j = 0$ occur at the same parameter, the bifurcation dynamics for the quartic nonlinear functional dynamical system is given by $\dot{x} = a_0(g(x) - a_1)^4$, as shown in Fig. 4.4. The bifurcation points are the *upper-* and *lower-saddle-node* bifurcations of the fourth-order. The fourth-order *upper-* and *lower-saddle-node* bifurcations for $a_0 > 0$, $dg/dx|_{x*} > 0$ and $a_0 < 0$, $dg/dx|_{x*} > 0$ are presented in Figs. 4.4a and b, respectively. The fourth-order *saddle-node* bi-

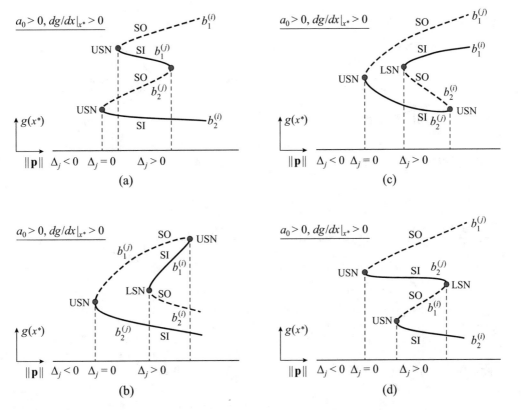

Figure 4.5: Stability and bifurcation of functional equilibriums in the 1-dimensional, quartic nonlinear functional dynamical system ($a_0 > 0$, $dg/dx|_{x^*} > 0$): (a) $b_2^{(i)} = b_1^{(j)}$, (b) $b_1^{(i)} = b_1^{(j)}$, (c) $b_2^{(i)} = b_2^{(j)}$, (d) $b_1^{(i)} = b_2^{(j)}$. LSN: *lower-saddle-node*, USN: *upper-saddle-node*, SI: sink, SO: source. Stable and unstable equilibriums are represented by solid and dashed curves, respectively. The bifurcation points are marked by circular symbols.

furcation possesses four branches rather than two branches for the second-order *saddle-node* bifurcation. For $B_i \leq B_j$, the fourth-order *upper-* and *lower-saddle-node* bifurcations are presented in Figs. 4.4c,d for $a_0 > 0$, $dg/dx|_{x^*} > 0$ and $a_0 < 0$, $dg/dx|_{x^*} > 0$.

If $\Delta_i = 0$ ($i \in \{1, 2\}$) occurs for new equilibrium appearance or vanishing only, the open loop of the bifurcation diagrams for $a_0 > 0$, $dg/dx|_{x^*} > 0$ is presented in Fig. 4.5. There are four cases for $a_0 > 0$ and $dg/dx|_{x^*} > 0$: (a) $b_2^{(i)} = b_1^{(j)}$, (b) $b_1^{(i)} = b_1^{(j)}$, (c) $b_2^{(i)} = b_2^{(j)}$, (d) $b_1^{(i)} = b_2^{(j)}$. The bifurcation points are the functional *upper-* and *lower-saddle-node* bifurcations of the second-order. In Figs. 4.6a–d, the functional *upper-* and *lower-saddle-node* bifurcations of the second-order in the open loop of bifurcation diagrams for $a_0 < 0$ and $dg/dx|_{x^*} > 0$

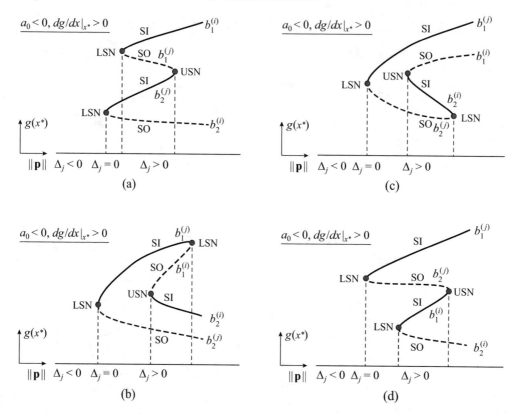

Figure 4.6: Open loops for stability and bifurcation of functional equilibriums in the 1-dimensional, quartic nonlinear functional dynamical system ($a_0 < 0$, $dg/dx|_{x*} > 0$): (a) $b_2^{(i)} = b_1^{(j)}$, (b) $b_1^{(j)} = b_1^{(i)} = b_2^{(i)} = -\frac{1}{2}B_i$. (c) $b_1^{(i)} = b_1^{(j)}$, (d) $b_2^{(i)} = b_2^{(j)}$, (e) $b_2^{(j)} = b_1^{(i)} = b_2^{(i)} = -\frac{1}{2}B_i$, (f) $b_1^{(i)} = b_2^{(j)}$. LSN: *lower-saddle-node*, USN: *upper-saddle-node*, SI: *sink*, SO: *source*. Stable and unstable equilibriums are represented by solid and dashed curves, respectively. The bifurcation points are marked by circular symbols.

are presented. Such a diagram of stability and bifurcation possesses three *saddle-node* bifurcations of the second order.

4.2   HIGHER-ORDER FUNCTIONAL EQUILIBRIUMS

Definition 4.3   Consider a 1-dimensional, quartic nonlinear functional dynamical system

$$\dot{x} = A(\mathbf{p})(g(x))^4 + B(\mathbf{p})(g(x))^3 + C(\mathbf{p})(g(x))^2 + D(\mathbf{p})g(x) + E(\mathbf{p})$$
$$= a_0(g(x) - b_1)^2[(g(x))^2 + B_2g(x) + C_2], \qquad (4.56)$$

where $A(\mathbf{p}) \neq 0$, and

$$\mathbf{p} = (p_1, p_2, \ldots, p_m)^{\mathrm{T}}. \tag{4.57}$$

(i) If

$$\Delta_2 = B_2^2 - 4C_2 < 0, \tag{4.58}$$

the corresponding standard functional form is

$$\dot{x} = a_0(g(x) - b_1)^2[(g(x) + \frac{1}{2}B_2)^2 + \frac{1}{4}(-\Delta_2)] \tag{4.59}$$

with a double-repeated functional equilibrium set of

$$x^* \in S_1 = \{a_1^{(s)} | g(a_1^{(s)}) = a_1, \ s = 1, 2, \ldots, N_1\} \cup \{\emptyset\}, \tag{4.60}$$
$$g(x^*) = a_1 = b_1.$$

   (a) For $a_0 > 0$, the flow is called a functional *upper-saddle* (US) flow.

   (b) For $a_0 < 0$, the flow is called a functional *lower-saddle* (LS) flow.

(ii) If

$$\Delta_2 = B_2^2 - 4C_2 > 0, \tag{4.61}$$

the quartic nonlinear functional dynamical system has two more functional equilibriums as

$$g(x^*) = b_1^{(2)} = -\frac{1}{2}(B_2 + \sqrt{\Delta_2}), \ g(x^*) = b_2^{(2)} = -\frac{1}{2}(B_2 - \sqrt{\Delta_2}). \tag{4.62}$$

(ii$_1$) The corresponding standard functional form is

$$\dot{x} = a_0(g(x) - a_1)^2(g(x) - a_2)(g(x) - a_3), \tag{4.63}$$

   where

$$x^* \in S_i = \{a_i^{(s)} | g(a_i^{(s)}) = a_i, \ s = 1, 2, \ldots, N_i\} \cup \{\emptyset\},$$
$$i = 1, 2, 3; \ a_1 \leq a_2 \leq a_3; \ a_1 = b_1 < \min\{b_1^{(2)}, b_2^{(2)}\}, \tag{4.64}$$
$$a_2 = \min\{b_1^{(2)}, b_2^{(2)}\}, \ a_3 = \max\{b_1^{(i)}, b_2^{(i)}\}.$$

   (ii$_{1a}$) For $a_0 > 0$ and $dg/dx|_{x^*} > 0$, the flow is called a functional (US:SI:SO) flow to (a_1, a_2, a_3).

   (ii$_{1b}$) For $a_0 > 0$ and $dg/dx|_{x^*} < 0$, the flow is called a functional (SI:SO:US) flow to (a_3, a_2, a_1).

   (ii$_{1c}$) For $a_0 < 0$ and $dg/dx|_{x^*} > 0$, the flow is called a functional (LS:SO:SI) flow to (a_1, a_2, a_3).

(ii_{1d}) For $a_0 < 0$ and $dg/dx|_{x^*} < 0$, the flow is called a functional (SO:SI:LS) flow to (a_3, a_2, a_1).

(ii_2) The corresponding standard functional form is

$$\dot{x} = a_0(g(x) - a_1)(g(x) - a_2)^2(g(x) - a_3), \tag{4.65}$$

where

$$x^* \in S_i = \{a_i^{(s)} | g(a_i^{(s)}) = a_i, \ s = 1, 2, \ldots, N_i\} \cup \{\emptyset\},$$
$$i = 1, 2, 3; \ a_1 \leq a_2 \leq a_3; \ a_1 = \min(b_1^{(2)}, b_2^{(2)}), \tag{4.66}$$
$$a_2 = b_1 > \min(b_1^{(2)}, b_2^{(2)}), \ a_3 = \max(b_1^{(i)}, b_2^{(i)}) > b_1.$$

(ii_{2a}) For $a_0 > 0$ and $dg/dx|_{x^*} > 0$, the flow is called a functional (SI:LS:SO) flow to (a_1, a_2, a_3).

(ii_{2b}) For $a_0 > 0$ and $dg/dx|_{x^*} < 0$, the flow is called a functional (SI:LS:SO) flow to (a_3, a_2, a_1).

(ii_{2c}) For $a_0 < 0$ and $dg/dx|_{x^*} > 0$, the flow is called a functional (SO:US:SI) flow to (a_1, a_2, a_3).

(ii_{2d}) For $a_0 < 0$ and $dg/dx|_{x^*} < 0$, the flow is called a functional (SO:US:SI) flow to (a_3, a_2, a_1).

(ii_3) The corresponding standard functional form is

$$\dot{x} = a_0(g(x) - a_1)(g(x) - a_2)(g(x) - a_3)^2, \tag{4.67}$$

where

$$x^* \in S_i = \{a_i^{(s)} | g(a_i^{(s)}) = a_i, \ s = 1, 2, \ldots, N_i\} \cup \{\emptyset\},$$
$$i = 1, 2, 3; \ a_1 \leq a_2 \leq a_3; \ a_1 = \min(b_1^{(2)}, b_2^{(2)}), \tag{4.68}$$
$$a_2 = \max(b_1^{(i)}, b_2^{(i)}), \ a_3 = b_1 > \max(b_1^{(i)}, b_2^{(i)}).$$

(ii_{3a}) For $a_0 > 0$ and $dg/dx|_{x^*} > 0$, the flow is called a (SI:SO:US) flow to (a_1, a_2, a_3).

(ii_{3b}) For $a_0 > 0$ and $dg/dx|_{x^*} < 0$, the flow is called a (US:SI:SO) flow to (a_3, a_2, a_1).

(ii_{3c}) For $a_0 < 0$ and $dg/dx|_{x^*} > 0$, the flow is called a (SO:SI:LS) flow to (a_1, a_2, a_3).

(ii_{3d}) For $a_0 < 0$ and $dg/dx|_{x^*} < 0$, the flow is called a (LS:SO:SI) flow to (a_3, a_2, a_1).

(ii$_4$) The corresponding standard functional form is

$$\dot{x} = a_0(g(x) - a_1)^3(g(x) - a_2),\qquad(4.69)$$

where

$$x^* \in S_i = \{a_i^{(s)}|g(a_i^{(s)}) = a_i, \ s = 1, 2, \ldots, N_i\} \cup \{\emptyset\},$$
$$i = 1, 2; \ a_1 \le a_2; \ a_1 = b_1 = \min(b_1^{(2)}, b_2^{(2)}), \ a_2 = \max(b_1^{(i)}, b_2^{(i)}).\qquad(4.70)$$

(ii$_{4a}$) For $a_0 > 0$ and $dg/dx|_{x^*} > 0$, the flow is called a functional (3rd SI:SO) flow. The bifurcation for functional (US:SI:SO)-equilibriums to functional (SI:LS:SO)-equilibrium to (a_1, a_2, a_3) is called the functional sink bifurcation of third order.

(ii$_{4b}$) For $a_0 > 0$ and $dg/dx|_{x^*} < 0$, the flow is called a functional (SI: 3rd SO) flow. The bifurcation for functional (SI:SO:US)-equilibrium to functional (SI:LS:SO)-equilibrium to (a_3, a_2, a_1) is called the functional source bifurcation of the third order.

(ii$_{4c}$) For $a_0 < 0$ and $dg/dx|_{x^*} > 0$, the flow is called a functional (3rd SO:SI) flow. The bifurcation for functional (LS:SO:SI)-equilibrium to functional (SO:US:SI)-equilibrium to (a_1, a_2, a_3) is called the functional source bifurcation of the third order.

(ii$_{4d}$) For $a_0 < 0$ and $dg/dx|_{x^*} < 0$, the flow is called a functional (SO:3rd SI) flow. The bifurcation for functional (SO:SI:LS)-equilibrium to functional (SO:US:SI)-equilibrium to (a_3, a_2, a_1) is called the functional sink bifurcation of the third order.

(ii$_5$) The corresponding standard functional form is

$$\dot{x} = a_0(g(x) - a_1)(g(x) - a_2)^3,\qquad(4.71)$$

where

$$x^* \in S_i = \{a_i^{(s)}|g(a_i^{(s)}) = a_i, \ s = 1, 2, \ldots, N_i\} \cup \{\emptyset\},$$
$$i = 1, 2; \ a_1 \le a_2; \ a_1 = \min(b_1^{(2)}, b_2^{(2)}), \ a_2 = b_1 = \max(b_1^{(i)}, b_2^{(i)}).\qquad(4.72)$$

(ii$_{5a}$) For $a_0 > 0$ and $dg/dx|_{x^*} > 0$, the flow is called a functional (SI:3rdSO) flow. The bifurcation of functional (SI:SO:US)-equilibriums to functional (SI:LS:SO)-equilibrium to (a_1, a_2, a_3) is called the functional source bifurcation of third order.

(ii$_{5b}$) For $a_0 > 0$ and $dg/dx|_{x^*} < 0$, the flow is called a functional (3rdSI:SO) flow. The bifurcation for functional (US:SI:SO)-equilibriums to functional (SI:US:SO)-equilibriums to (a_3, a_2, a_1) is called the functional source bifurcation of the third order.

(ii$_{5c}$) For $a_0 < 0$ and $dg/dx|_{x^*} > 0$, the flow is called a functional (SO:3rdSI) flow. The bifurcation for functional (SO:SI:LS)-equilibriums to functional (SO:US:SI)-equilibriums to (a_1, a_2, a_3) is called the functional sink bifurcation of the third order.

(ii$_{5d}$) For $a_0 < 0$ and $dg/dx|_{x^*} < 0$, the flow is called a functional (3rdSO:SI) flow. The bifurcation for functional (LS:SO:SI)-equilibriums to functional (SO:US:SI)-equilibriums to (a_3, a_2, a_1) is called the functional sink bifurcation of the third order.

(iii) If

$$\Delta_2 = B_2^2 - 4C_2 = 0, \tag{4.73}$$

the quartic nonlinear functional dynamical system has one more extra double-repeated equilibrium as

$$g(x^*) = b_1^{(2)} = b_2^{(2)} = b_2 = -\frac{1}{2}B_2. \tag{4.74}$$

(iii$_1$) The corresponding standard functional form for $b_2 > b_1$ is

$$\dot{x} = a_0(g(x) - b_1)^2(g(x) - b_2)^2, \tag{4.75}$$

where

$$x^* \in S_i = \{a_i^{(s)} | g(a_i^{(s)}) = a_i, \; s = 1, 2, \ldots, N_i\} \cup \{\emptyset\}, \\ i = 1, 2; \; a_1 \leq a_2; \; a_1 = b_1, \; a_2 = b_2. \tag{4.76}$$

(iii$_{1a}$) For $a_0 > 0$ and $dg/dx|_{x^*} > 0$, the flow is called a functional (US:US) flow. The bifurcation for the functional (US:SI:SO)-equilibriums appearance to (a_1, a_2, a_3) is called the functional *upper-saddle* bifurcation of the second order.

(iii$_{1b}$) For $a_0 > 0$ and $dg/dx|_{x^*} < 0$, the flow is called a functional (US:US) flow. The bifurcation for the functional (SI:SO:US)-equilibriums appearance to (a_3, a_2, a_1) is called the functional *upper-saddle* bifurcation of the second order.

(iii$_{1c}$) For $a_0 < 0$ and $dg/dx|_{x^*} > 0$, the flow is called a (LS:LS) flow. The bifurcation for functional (LS:SO:SI)-equilibriums appearance to (a_1, a_2, a_3) is called the functional *lower-saddle* bifurcation of the second order.

(iii$_{1d}$) For $a_0 < 0$ and $dg/dx|_{x^*} < 0$, the flow is called a (LS:LS) flow. The bifurcation for functional (SO:SI:LS)-equilibriums appearance to (a_3, a_2, a_1) is called the functional *lower-saddle* bifurcation of the second order.

(iii$_2$) The corresponding standard functional form for $b_1 > b_2$ is

$$\dot{x} = a_0(g(x) - b_2)^2(g(x) - b_1)^2, \tag{4.77}$$

where

$$x^* \in S_i = \{a_i^{(s)} | g(a_i^{(s)}) = a_i, \ s = 1, 2, \dots, N_i\} \cup \{\emptyset\},$$

$$i = 1, 2; \ a_1 \leq a_2; \ a_1 = b_2, \ a_2 = b_1.$$

(4.78)

(iii$_{2a}$) For $a_0 > 0$ and $dg/dx|_{x^*} > 0$, the flow is called a functional (US:US) flow. The bifurcation of functional (SI:SO:US)-equilibrium appearance to (a_1, a_2, a_3) is called the functional *upper-saddle* bifurcation of the second order.

(iii$_{2b}$) For $a_0 > 0$ and $dg/dx|_{x^*} < 0$, the flow is called a functional (US:US) flow. The bifurcation for functional (US:SI:SO)-equilibriums appearance to (a_3, a_2, a_1) is called the functional *upper-saddle* bifurcation of the second order.

(iii$_{2c}$) For $a_0 < 0$ and $dg/dx|_{x^*} > 0$, the flow is called a functional (LS:LS) flow. The bifurcation for functional (SO:SI:LS)-equilibriums appearance to (a_1, a_2, a_3) is called the functional *lower-saddle* bifurcation of the second order.

(iii$_{2d}$) For $a_0 < 0$ and $dg/dx|_{x^*} < 0$, the flow is called a functional (LS:LS) flow. The bifurcation for functional (LS:SO:SI)-equilibriums appearance to (a_3, a_2, a_1) is called the functional *lower-saddle* bifurcation of the second order.

(iii$_3$) The corresponding standard functional form with $b_1 = b_2 = a_1$ is

$$\dot{x} = a_0(x - a_1)^4,$$

(4.79)

where

$$x^* \in S_1 = \{a_1^{(s)} | g(a_1^{(s)}) = a_1, \ s = 1, 2, \dots, N\} \cup \{\emptyset\},$$

$$a_1 = b_2 = b_1.$$

(4.80)

(iii$_{3a}$) For $a_0 > 0$ and $dg/dx|_{x^*} > 0$, the flow is called a fourth-order functional *upper-saddle* (US) flow. The bifurcation for the functional *upper-saddle* (US) equilibrium to functional (SI:LS:SO)-equilibriums to (a_1, a_2, a_3) is called the functional *upper-saddle-node* bifurcation of the fourth order.

(iii$_{3b}$) For $a_0 > 0$ and $dg/dx|_{x^*} < 0$, the flow is called a fourth-order functional *upper-saddle* (US) flow. The bifurcation for the functional *upper-saddle* (US) equilibrium to functional (SI:LS:SO)-equilibriums to (a_3, a_2, a_1) is called the functional *upper-saddle-node* bifurcation of the fourth order.

(iii$_{3c}$) For $a_0 < 0$ and $dg/dx|_{x^*} > 0$, the flow is called a fourth-order *lower-saddle* (LS) flow. The bifurcation for a functional *lower-saddle* (LS) equilibrium to functional (SO:US:SI)-equilibrium to (a_1, a_2, a_3) is called the functional *lower-saddle* bifurcation of the fourth order.

(iii$_{3d}$) For $a_0 < 0$ and $dg/dx|_{x^*} < 0$, the flow is called a fourth-order *lower-saddle* (LS) flow. The bifurcation for the functional *lower-saddle* (LS) equilibrium to functional (SO:US:SI)-equilibriums to (a_3, a_2, a_1) is called the functional *lower-saddle* bifurcation of the fourth order.

Definition 4.4   Consider a 1-dimensional, quartic nonlinear functional dynamical system

$$
\begin{aligned}
\dot{x} &= A(\mathbf{p})(g(x))^4 + B(\mathbf{p})(g(x))^3 + C(\mathbf{p})(g(x))^2 + D(\mathbf{p})g(x) + E(\mathbf{p}) \\
&= a_0(\mathbf{p})[(g(x))^2 + B_1 g(x) + C_1]^2,
\end{aligned}
\tag{4.81}
$$

where $A(\mathbf{p}) \neq 0$, and

$$
\mathbf{p} = (p_1, p_2, \ldots, p_m)^{\mathrm{T}}.
\tag{4.82}
$$

(i) If

$$
\Delta_1 = B_1^2 - 4C_1 < 0,
\tag{4.83}
$$

the quartic nonlinear functional dynamical system does not have any equilibrium.

(i$_1$) For $a_0 > 0$, the non-equilibrium flow is called a positive flow.

(i$_2$) For $a_0 < 0$, the non-equilibrium flow is called a negative flow.

(ii) If

$$
\Delta_1 = B_1^2 - 4C_1 > 0,
\tag{4.84}
$$

the 1-dimensional quartic nonlinear functional dynamical system has two double-repeated functional equilibriums as

$$
g(x^*) = b_1^{(1)} = -\frac{1}{2}(B_1 + \sqrt{\Delta_1}), \ \ g(x^*) = b_2^{(1)} = -\frac{1}{2}(B_1 - \sqrt{\Delta_1}).
\tag{4.85}
$$

The corresponding standard functional form is

$$
\dot{x} = a_0(g(x) - a_1)^2(g(x) - a_2)^2,
\tag{4.86}
$$

where

$$
\begin{aligned}
x^* &\in S_i = \{a_i^{(s)} | g(a_i^{(s)}) = a_i, \ s = 1, 2, \ldots, N_i\} \cup \{\emptyset\}, \\
i &= 1, 2; \ a_1 \leq a_2; \ a_1 = \min\{b_1^{(1)}, b_2^{(1)}\}, \ a_2 = \max\{b_1^{(1)}, b_2^{(1)}\}.
\end{aligned}
\tag{4.87}
$$

(ii$_1$) For $a_0 > 0$ and $dg/dx|_{x^*} \neq 0$, the flow is called a functional (US:US) flow.

(ii$_2$) For $a_0 < 0$ and $dg/dx|_{x^*} \neq 0$, the flow is called a functional (LS:LS) flow.

(iii) If

$$\Delta_1 = B_1^2 - 4C_1 = 0, \tag{4.88}$$

the 1-dimensional quartic nonlinear functional dynamical system has one 4-time repeated equilibrium as

$$g(x^*) = b_1^{(1)} = b_2^{(1)} = -\frac{1}{2}B_1. \tag{4.89}$$

The corresponding standard functional form is

$$\dot{x} = a_0(g(x) - a_1)^4, \tag{4.90}$$

where

$$
\begin{aligned}
&x^* \in S_1 = \{a_1^{(s)} | g(a_1^{(s)}) = a_1, \ s = 1, 2, \ldots, N_1\} \cup \{\emptyset\}, \\
&a_1 = b_1 = b_1^{(1)} = b_2^{(1)}.
\end{aligned}
\tag{4.91}
$$

(iii$_1$) For $a_0 > 0$ and $dg/dx|_{x^*} \neq 0$, the flow is called a fourth-order functional US flow. The bifurcation for functional (US:US)-equilibriums appearance is called the functional *upper-saddle* bifurcation of the fourth order.

(iii$_2$) For $a_0 < 0$ and $dg/dx|_{x^*} \neq 0$, the flow is called a fourth-order functional LS flow. The bifurcation for the functional (LS:LS)-equilibriums appearance is called the functional *lower-saddle* bifurcation of the fourth order.

From a 1-dimensional, quartic nonlinear functional dynamical system with singularity, the saddle equilibrium with and without intersection with simple functional equilibriums are presented in Figs. 4.7 and 4.8. In Figs. 4.7a,d, the functional *upper-saddle* equilibrium for $a_0 > 0$ and $dg/dx|_{x^*} > 0$ does not intersect with any branch of the simple equilibriums. In Figs. 4.7b,c, the functional *upper-saddle* equilibrium for $a_0 > 0$ and $dg/dx|_{x^*} > 0$ intersects with one branch of the simple functional equilibriums, and the functional *upper-saddle* equilibrium switches to the functional *lower-saddle* equilibrium with functional source and sink equilibriums, which are called the functional source and sink bifurcations of the third-order, accordingly. In Fig. 4.7e, the functional *upper-saddle* equilibrium for $a_0 > 0$ and $dg/dx|_{x^*} > 0$ intersects with a double-repeated equilibrium with a functional *upper-saddle*. The intersected point is an unstable functional equilibrium, which is called a fourth-order functional *upper-saddle-node* bifurcation. In Fig. 4.7f, the two second-order functional *upper-saddle* equilibriums are presented for $a_0 > 0$ and $dg/dx|_{x^*} \neq 0$. The two functional *upper-saddle* equilibriums appear at the bifurcation of the fourth-order functional *upper-saddle* bifurcation.

Similarly, the functional *lower-saddle* equilibrium for $a_0 < 0$ and $dg/dx|_{x^*} > 0$ does not intersect with any branch of the simple equilibriums, as shown in Figs. 4.8a,d. In Figs. 4.8b,c, the functional *lower-saddle* equilibrium for $a_0 < 0$ and $dg/dx|_{x^*} > 0$ intersects with one branch of the simple functional equilibriums, and the functional *lower-saddle* equilibrium switches to

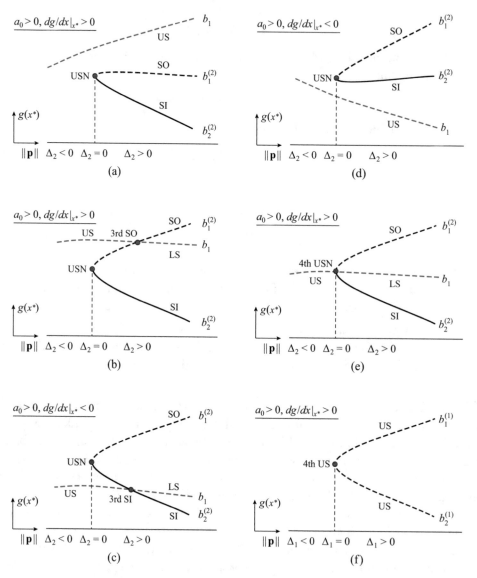

Figure 4.7: Three equilibriums with and without intersection in the 1-dimensional, quartic nonlinear functional dynamical system ($a_0 > 0$): (a) without intersection $b_1 > b_1^{(2)}$ ($dg/dx|_{x*} > 0$), (b) an intersection at $b_1 = b_1^{(2)}$ ($dg/dx|_{x*} > 0$), (c) an intersection at $b_1 = b_2^{(2)}$ ($dg/dx|_{x*} < 0$), (d) without intersection $b_1 < b_2^{(2)}$ ($dg/dx|_{x*} < 0$), (e) an intersection at $b_1 = -\frac{1}{2}B_2$ ($dg/dx|_{x*} > 0$), (f) $\Delta_1 = 0$ ($dg/dx|_{x*} \neq 0$). LSN: *lower-saddle-node*, USN: *upper-saddle-node*, SI: sink, SO: source, US: *upper-saddle*, LS: *lower-saddle*. Stable and unstable equilibriums are represented by solid and dashed curves, respectively. The bifurcation points are marked by circular symbols.

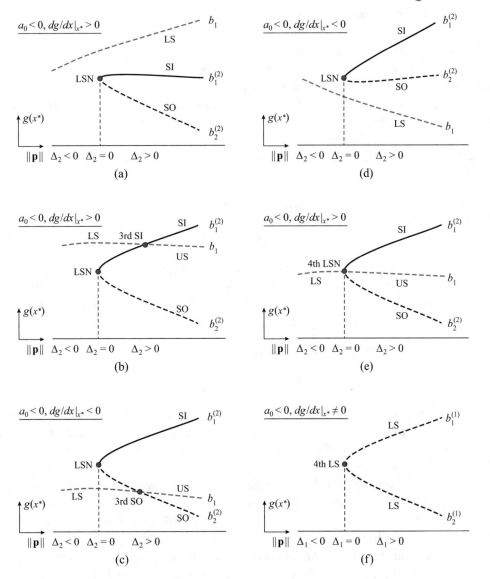

Figure 4.8: Three equilibriums with and without intersection in the 1-dimensional, quartic nonlinear dynamical system ($a_0 < 0$): (a) without intersection $b_1 > b_1^{(2)} (dg/dx|_{x^*} > 0)$, (b) an intersection at $b_1 = b_1^{(2)} (dg/dx|_{x^*} > 0)$, (c) an intersection at $b_1 = b_2^{(2)} (dg/dx|_{x^*} < 0)$, (d) without intersection $b_1 < b_2^{(2)} (dg/dx|_{x^*} < 0)$, (e) an intersection at $b_1 = -\frac{1}{2}B_2$ ($dg/dx|_{x^*} > 0$), (f) $\Delta_1 = 0$ ($dg/dx|_{x^*} \neq 0$). LSN: *lower-saddle-node*, USN: *upper-saddle-node*, SI: *sink*, SO: *source*, US: *upper-saddle*, LU: *lower-saddle*. Stable and unstable equilibriums are represented by solid and dashed curves, respectively. The bifurcation points are marked by circular symbols.

the functional *upper-saddle* equilibrium with functional source and sink equilibriums, which are called the functional source and sink bifurcations of the third order, accordingly. In Fig. 4.8e, the functional *lower-saddle* equilibrium for $a_0 < 0$ and $dg/dx|_{x*} > 0$ intersects with a repeated equilibrium with a functional *lower-saddle*. The intersection point is an unstable functional equilibrium, which is called a fourth-order functional *lower-saddle-node* bifurcation. In Fig. 4.8f, the two second-order functional *lower-saddle* equilibriums are presented for $a_0 < 0$ and $dg/dx|_{x*} \neq 0$. The two functional *lower-saddle* equilibriums appear at the fourth-order functional *lower-saddle* bifurcation.

Consider a 1-dimensional, quartic nonlinear functional dynamical system with two double equilibriums.

(i) For $b \neq a$, the functional dynamical system is

$$\dot{x} = a_0(\mathbf{p})(g(x) - b(\mathbf{p}))^2 (g(x) - a(\mathbf{p}))^2. \tag{4.92}$$

For such a functional system, if $a_0 > 0$ and $dg/dx|_{x*} \neq 0$, two double-repeated functional equilibriums of $g(x^*) = a, b$ are two functional *upper-saddles*, which are unstable. If $a_0 < 0$ and $dg/dx|_{x*} \neq 0$, two double-repeated functional equilibriums of $g(x^*) = a, b$ are two *lower-saddles*, which are unstable.

(ii) For $b = a$, the dynamical functional system on the boundary is

$$\dot{x} = a_0(\mathbf{p})(g(x) - b(\mathbf{p}))^4. \tag{4.93}$$

With parameter changes, the bifurcation diagram for the quartic nonlinear functional system is presented in Fig. 4.9. Stable and unstable functional equilibriums are represented by solid and dashed curves, respectively. The bifurcation point is marked by a circular symbol. In Fig. 4.9a, if $a_0 > 0$ and $dg/dx|_{x*} \neq 0$, two double-repeated functional equilibriums of $g(x^*) = a, b$ are the functional *upper-saddles* of the second-order. The two functional *upper-saddles* intersect at a point of $g(x^*) = a = b$ with the fourth-order multiplicity, which is a functional *upper-saddle* bifurcation of the fourth order for the (US:US) to (US:US) equilibriums. If $a_0 < 0$ and $dg/dx|_{x*} \neq 0$, two double-repeated functional equilibriums of $g(x^*) = a, b$ are the functional *lower-saddle* of the second order, which are intersected at a point of $x^* = a = b$, as shown in Fig. 4.9b. Such a functional equilibrium with the fourth-order multiplicity is called a functional *lower-saddle* bifurcation of the fourth order for the (LS:LS) to (LS:LS) equilibrium.

To illustrate the stability and bifurcation of functional equilibrium with singularity in a 1-dimensional, quadratic nonlinear functional system, the function equilibrium of $\dot{x} = a_0(g(x) - a_1)^4$ is presented in Fig. 4.10. The fourth-order, functional *upper-* and *lower-saddle* equilibriums of $x^* = a_1$ with the fourth-order multiplicity are unstable, and the functional upper- and *lower-saddle* equilibriums of the fourth-order are invariant. At $a_0 = 0$, the functional *lower-saddle*

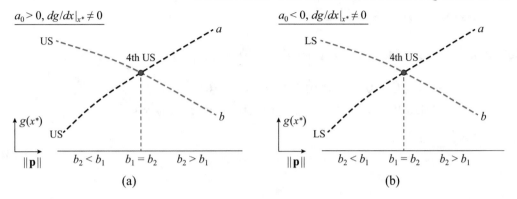

Figure 4.9: Stability and bifurcation of two functional US or LS equilibriums with intersection in the 1-dimensional, quartic nonlinear functional dynamical system: (a) functional (US:US)-flow ($a_0 > 0$, $dg/dx|_{x*} \neq 0$), (b) functional (LS:LS)-flow ($a_0 < 0$, $dg/dx|_{x*} \neq 0$). 4th LS: fourth-order *lower-saddle* bifurcation, 4th US: fourth-order *upper-saddle* bifurcation. Stable and unstable equilibriums are represented by solid and dashed curves, respectively. The bifurcation points are marked by circular symbols.

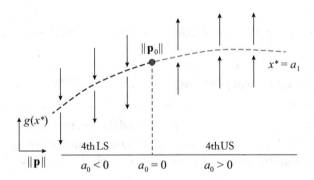

Figure 4.10: Stability of a repeated equilibrium with the fourth multiplicity in the 1-dimensional, quartic nonlinear dynamical system: Stable and unstable equilibriums are represented by solid and dashed curves, respectively. The stability switching is labeled by a circular symbol.

equilibrium switches to the *upper-saddle* equilibrium, which is a switching point marked by a circular symbol.

For further discussion on the switching bifurcations in the quartic functional system, the following definitions are presented.

4.3   SWITCHING BIFURCATIONS

Definition 4.5   Consider a 1-dimensional, quartic nonlinear functional dynamical system

$$\dot{x} = A(\mathbf{p})(g(x))^4 + B(\mathbf{p})(g(x))^3 + C(\mathbf{p})(g(x))^2 + D(\mathbf{p})g(x) + E(\mathbf{p})$$
$$= a_0(\mathbf{p})(g(x) - a)(g(x) - b)[(g(x))^2 + B_2(\mathbf{p})g(x) + C_2(\mathbf{p})], \tag{4.94}$$

where $A(\mathbf{p}) \neq 0$, and

$$\mathbf{p} = (p_1, p_2, \ldots, p_m)^{\mathrm{T}}. \tag{4.95}$$

(i) If

$$\Delta_2 = B_2^2 - 4C_2 < 0, \ \{a_1, a_2\} = \text{sort}\{a, b\}, \ a_1 \leq a_2,$$
$$x^* \in S_i = \{a_i^{(s)} | g(a_i^{(s)}) = a_i, \ s = 1, 2, \ldots, N_i\} \cup \{\emptyset\}, \tag{4.96}$$
$$i = 1, 2; \ a_1 \leq a_2;$$

the quartic nonlinear functional dynamical system has two equilibriums. The standard functional form is

$$\dot{x} = a_0(\mathbf{p})(g(x) - a_1)(g(x) - a_2)[(g(x) + \frac{1}{2}B_2)^2 + \frac{1}{4}(-\Delta_2)]. \tag{4.97}$$

(i$_1$) For $a_0 dg/dx|_{x^*} > 0$, the functional equilibrium flow is a (SI:SO) flow. The functional equilibrium of $g(x^*) = a_1$ is stable (sink, $df/dx|_{x^*} < 0$) and the functional equilibrium of $g(x^*) = a_2$ is unstable (source, $df/dx|_{x^*} > 0$).

(i$_2$) For $a_0 dg/dx|_{x^*} < 0$, the functional equilibrium flow is a (SO:SI) flow. The functional equilibrium of $g(x^*) = a_1$ is unstable (source, $df/dx|_{x^*} > 0$) and the functional equilibrium of $g(x^*) = a_2$ is stable (sink, $df/dx|_{x^*} < 0$).

(i$_3$) Under

$$\Delta_1 = (a_1 - a_2)^2 = 0 \text{ with } a_1 = a_2 \tag{4.98}$$

the quartic nonlinear functional dynamical system has a standard form as

$$\dot{x} = f(x, \mathbf{p}) = a_0(g(x) - a_1)^2[(g(x) + \frac{1}{2}B_2)^2 + \frac{1}{4}(-\Delta_2)]. \tag{4.99}$$

(i$_{3a}$) For $a_0(\mathbf{p}) > 0$, the functional equilibrium of $x^* = a_1^{(s)} \in S_1$ with $g(x^*) = a_1$ is unstable (a functional *upper-saddle* of second-order, $d^2 f/dx^2|_{x^*=a_1} > 0$). Such a flow is called a functional *upper-saddle* flow. The bifurcation of functional equilibrium of $x^* = a_1^{(s)} \in S_1$ with $g(x^*) = a_1$ for two functional equilibriums switching of $g(x^*) = a_1, a_2$ is called a functional *upper-saddle-node* bifurcation of the second order at a point $\mathbf{p} = \mathbf{p}_1$.

(i_{3b}) For $a_0(\mathbf{p}) < 0$, the functional equilibrium of $x^* = a_1^{(s)} \in S_1$ with $g(x^*) = a_1$ is unstable (a *lower-saddle* of second order, $d^2 f/dx^2|_{x*} < 0$). Such a flow is called a functional *lower-saddle* flow. The bifurcation of functional equilibrium at of $x^* = a_1^{(s)} \in S_1$ with $g(x^*) = a_1$ for two functional equilibriums switching of $g(x^*) = a_1, a_2$ is called a functional *lower-saddle-node* bifurcation of the second order at a point $\mathbf{p} = \mathbf{p}_1$.

(ii) If

$$\Delta_2 = B_2^2 - 4C_2 > 0, \tag{4.100}$$

the 1-dimensional quartic nonlinear functional dynamical system has four functional equilibriums as

$$x^* \in S_i = \{a_i^{(s)}|g(a_i^{(s)}) = a_i, \ s = 1, 2, \ldots, N_i\} \cup \{\emptyset\},$$

$$i \in \{1, 2, 3, 4\}; \ \{a_1, a_2, a_3, a_4\} = \text{sort}\{a, b, b_1^{(2)}, b_2^{(2)}\}, \ a_i < a_{i+1} \tag{4.101}$$

$$b_1^{(2)} = -\frac{1}{2}(B_2 + \sqrt{\Delta_2}), \ b_2^{(2)} = -\frac{1}{2}(B_2 - \sqrt{\Delta_2}).$$

The corresponding standard functional form is

$$\dot{x} = a_0(g(x) - a_1)(g(x) - a_2)(g(x) - a_3)(g(x) - a_4). \tag{4.102}$$

(ii_1) For $a_0 dg/dx|_{x*} > 0$, the flow is called a functional (SI:SO:SI:SO) flow.

(ii_2) For $a_0 dg/dx|_{x*} < 0 < 0$, the flow is called a functional (SO:SI:SO:SO) flow.

(ii_3) Under

$$x^* \in S_i = \{a_i^{(s)}|g(a_i^{(s)}) = a_i, \ s = 1, 2, \ldots, N_i\} \cup \{\emptyset\},$$

$$i \in \{i_1, i_3, i_4\}; \ \Delta_{i_1 i_2} = (a_{i_1} - a_{i_2})^2 = 0, \tag{4.103}$$

$$a_{i_1} = a_{i_2}, \ i_1, i_2 \in \{1, 2, 3, 4\}, \ i_1 \neq i_2.$$

Then the standard form is

$$\dot{x} = f(x, \mathbf{p}) = a_0(x - a_{i_1})^2(x - a_{i_3})(x - a_{i_4})$$

$$i_\alpha \in \{1, 2, 3, 4\}, \ \alpha = 1, 3, 4. \tag{4.104}$$

(ii_{3a}) The functional equilibrium of $x^* = a_{i_1}^{(s)} \in S_{i_1}$ with $g(x^*) = a_{i_1}$ is unstable (a functional *upper-saddle* of second order, $d^2 f/dx^2|_{x*} > 0$). Such a flow is called a functional *upper-saddle* flow at $x^* = a_{i_1}^{(s)}$. The bifurcation of functional equilibrium at $x^* = a_{i_1}^{(s)}$ for two functional equilibriums switching of $g(x^*) = a_{i_1}, a_{i_2}$ is called an *upper-saddle-node* bifurcation of the second order at a point $\mathbf{p} = \mathbf{p}_1$.

(ii$_{3b}$) The functional equilibrium of $x^* = a_{i_1}^{(s)} \in S_{i_1}$ with $g(x^*) = a_{i_1}$ is unstable (a functional *lower-saddle* of the second-order, $d^2 f/dx^2|_{x^*} < 0$). Such a flow is called a functional *lower-saddle* flow at $x^* = a_{i_1}^{(s)}$. The bifurcation of functional equilibrium at $x^* = a_{i_1}^{(s)}$ for two functional equilibriums switching of $g(x^*) = a_{i_1}, a_{i_2}$ is called a *lower-saddle-node* bifurcation of the second-order at a point $\mathbf{p} = \mathbf{p}_1$.

(ii$_4$) Under

$$
\begin{aligned}
&x^* \in S_i = \{a_i^{(s)} | g(a_i^{(s)}) = a_i, \ s = 1, 2, \ldots, N_i\} \cup \{\emptyset\}, \\
&i = \{i_1, i_4\}; \ \Delta_{i_1 i_2} = (a_{i_1} - a_{i_2})^2 = 0, \ \Delta_{i_2 i_3} = (a_{i_2} - a_{i_3})^2 = 0 \quad (4.105) \\
&a_{i_1} = a_{i_2} = a_{i_3}, \ i_1, i_2, i_3 \in \{1, 2, 3, 4\}, \ i_1 \neq i_2 \neq i_3.
\end{aligned}
$$

Then the standard functional form is

$$
\begin{aligned}
\dot{x} &= f(x, \mathbf{p}) = a_0 (g(x) - a_{i_1})^3 (g(x) - a_{i_4}) \\
&i_\alpha \in \{1, 2, 3, 4\}, \ \alpha = 1, 4.
\end{aligned} \quad (4.106)
$$

(ii$_{4a}$) The functional equilibrium of $x^* = a_{i_1}^{(s)} \in S_{i_1}$ with $g(x^*) = a_{i_1}$ is unstable (a source of the third order, $d^3 f/dx^3|_{x^*} > 0$). Such a flow is called a third-order source flow at $x^* = a_{i_1}^{(s)}$. The bifurcation at functional equilibrium $x^* = a_{i_1}^{(s)}$ for three simple functional equilibriums switching of $g(x^*) = a_{i_1}, a_{i_2}, a_{i_3}$ is called a source bifurcation of the third order at a point $\mathbf{p} = \mathbf{p}_1$.

(ii$_{4b}$) The functional equilibrium of $x^* = a_{i_1}^{(s)} \in S_{i_1}$ with $g(x^*) = a_{i_1}$ is stable (a sink of the third order, $d^3 f/dx^3|_{x^*} < 0$). Such a flow is called a third-order sink flow at $x^* = a_{i_1}^{(s)}$. The bifurcation at functional equilibrium $x^* = a_{i_1}^{(s)}$ for three simple functional equilibriums switching of $g(x^*) = a_{i_1}, a_{i_2}, a_{i_3}$ is called a sink bifurcation of the third order at a point $\mathbf{p} = \mathbf{p}_1$.

(ii$_5$) Under

$$
\begin{aligned}
&x^* \in S_{i_1} = \{a_{i_1}^{(s)} | g(a_{i_1}^{(s)}) = a_{i_1}, \ s = 1, 2, \ldots, N_{i_1}\} \cup \{\emptyset\}, \\
&\Delta_{i_1, i_2} = (a_{i_1} - a_{i_2})^2 = 0, \ \Delta_{i_2 i_3} = (a_{i_2} - a_{i_3})^2 = 0, \\
&\Delta_{i_3, i_4} = (a_{i_4} - a_{i_4})^2 = 0, \ a_{i_1} = a_{i_2} = a_{i_3} = a_{i_4}, \\
&i_1, i_2, i_3, i_4 \in \{1, 2, 3, 4\}, \ i_1 \neq i_2 \neq i_3 \neq i_4.
\end{aligned} \quad (4.107)
$$

Then the standard form is

$$
\dot{x} = f(x, \mathbf{p}) = a_0 (g(x) - a_{i_1})^4. \quad (4.108)
$$

(ii$_{5a}$) The functional equilibrium of $x^* = a_{i_1}^{(s)} \in S_{i_1}$ with $g(x^*) = a_{i_1}$ is unstable (an *upper-saddle* of the fourth-order, $d^4 f/dx^4|_{x^* = a_{i_1}} > 0$). Such a flow is

called a fourth-order *upper-saddle* flow at $x^* = a_{i_1}^{(s)}$. The bifurcation of equilibrium at $x^* = a_{i_1}^{(s)}$ for four simple functional equilibriums bundle switching of $g(x^*) = a_{1,2,3,4}$ is called a functional *upper-saddle-node* bifurcation of the fourth order at a point $\mathbf{p} = \mathbf{p}_1$.

(ii$_{5b}$) The functional equilibrium of $x^* = a_{i_1}^{(s)} \in S_{i_1}$ with $g(x^*) = a_{i_1}$ is stable (a sink of the fourth-order, $d^4 f/dx^4|_{x^*} < 0$). Such a flow is called a fourth-order *lower-saddle* flow at $x^* = a_{i_1}$. The bifurcation of equilibrium at $x^* = a_{i_1}$ for four simple functional equilibriums bundle switching of $g(x^*) = a_{1,2,3,4}$ is called a functional *lower-saddle* bifurcation of the fourth order at a point $\mathbf{p} = \mathbf{p}_1$.

(iii) If

$$\Delta_2 = B_2^2 - 4C_2 = 0, \tag{4.109}$$

the 1-dimensional quartic nonlinear functional dynamical system has three equilibrium as

$$x^* \in S_i = \{a_i^{(s)} | g(a_i^{(s)}) = a_i, \ s = 1, 2, \ldots, N_i\} \cup \{\emptyset\},$$
$$i = \{i_1, i_2, i_3\}; \ \{a_1, a_2, a_3\} = \text{sort}\{a, b, b_1^{(2)} = b_2^{(2)}\}, \ a_i < a_{i+1} \tag{4.110}$$
$$a_{i_\alpha} \in \{a_1, a_2, a_3\}, \ \alpha \in \{1, 2, 3\}; \ b_1^{(2)} = b_2^{(2)} = -\frac{1}{2}B_2.$$

The corresponding standard functional form is

$$\dot{x} = a_0(g(x) - a_{i_1})^2(g(x) - a_{i_2})(g(x) - a_{i_3}). \tag{4.111}$$

(iii$_1$) The functional equilibrium of $x^* = a_{i_1}^{(s)} \in S_{i_1}$ with $g(x^*) = a_{i_1}$ is unstable (a functional *upper-saddle*, $d^2 f/dx^2|_{x^*=a_1} > 0$), the flow is a functional upper flow at $x^* = a_{i_1}^{(s)}$. The bifurcation of equilibrium at $x^* = a_{i_1}^{(s)}$ for the appearing or vanishing of two simple functional equilibriums is called the functional *upper-saddle-node* bifurcation of the second order.

(iii$_2$) The functional equilibrium of $x^* = a_{i_1}^{(s)} \in S_{i_1}$ with $g(x^*) = a_{i_1}$ is unstable (a functional *lower-saddle*, $d^2 f/dx^2|_{x^*} < 0$), the flow is a functional *lower-saddle* flow at $x^* = a_{i_1}^{(s)}$. The bifurcation of equilibrium at $x^* = a_{i_1}^{(s)}$ for the appearing or vanishing of two simple functional equilibriums is called the functional *lower-saddle-node* bifurcation of the second order.

(iii$_3$) Under

$$x^* \in S_i = \{a_i^{(s)} | g(a_i^{(s)}) = a_i, \ s = 1, 2, \ldots, N_i\} \cup \{\emptyset\},$$
$$i = \{i_1, i_3\}; \ \Delta_{i_1 i_2} = (a_{i_1} - a_{i_2})^2 = 0, \ a_{i_1} = a_{i_2}, \ a_{i_1} \neq a_{i_3}, \tag{4.112}$$
$$i_1, i_2, i_3 \in \{1, 2, 3\}, \ i_1 \neq i_2 \neq i_3,$$

the standard form is

$$\dot{x} = f(x, \mathbf{p}) = a_0(g(x) - a_{i_1})^2(g(x) - a_{i_3})^2$$
$$i_\alpha \in \{1, 2\}, \ \alpha = 1, 3.$$

(4.113)

The functional equilibriums of $x^* = a_i^{(s)} \in S_i$ with $g(x^*) = a_i$ $(i = i_1, i_3)$ are unstable (functional *upper-saddles* of the second order, $d^2 f/dx^2|_{x^*} > 0$) and unstable (functional *lower-saddle* of the second order, $d^2 f/dx^2|_{x^*} < 0$). Such a flow is called a functional (US:US) or (LS:LS) flow. The bifurcation of functional equilibrium at $x^* = a_{i_1}^{(s)}$ for two functional equilibriums switching of $g(x^*) = a_{i_1}, a_{i_2}$ and at $x^* = a_{i_3}^{(s)}$ for onset of two functional equilibriums $g(x^*) = a_{i_3}, a_{i_4}$ is called a functional (US:US) or (LS:LS) at a point $\mathbf{p} = \mathbf{p}_1$.

(iii$_4$) Under

$$\Delta_{i_1 i_2} = (a_{i_1} - a_{i_2})^2 = 0, \ a_{i_1} = a_{i_2}, \ a_{i_1} = a_{i_3},$$
$$i_1, i_2, i_3 \in \{1, 2, 3\}, \ i_1 \neq i_2 \neq i_3,$$
$$x^* \in S_{i_1} = \{a_{i_1}^{(s)}|g(a_{i_1}^{(s)}) = a_{i_1}, \ s = 1, 2, \ldots, N_{i_1}\} \cup \{\emptyset\},$$

(4.114)

the standard form is

$$\dot{x} = f(x, \mathbf{p}) = a_0(g(x) - a_{i_1})^4.$$

(4.115)

(iii$_{4a}$) For $a_0 > 0$ and $dg/dx|_{x^*} \neq 0$, the functional equilibrium of $x^* = a_{i_1}^{(s)} \in S_{i_1}$ with $g(x^*) = a_{i_1}$ is unstable (a functional *upper-saddle* of the fourth-order, $d^4 f/dx^4|_{x^*} > 0$). Such a flow is called a functional *upper-saddle* flow of the fourth-order. The bifurcation of equilibrium at $x^* = a_{i_1}^{(s)}$ for two simple functional equilibriums switching to four simple functional equilibriums is called a fourth-order *upper-saddle-node* at a point $\mathbf{p} = \mathbf{p}_1$.

(iii$_{4b}$) For $a_0 < 0$ and $dg/dx|_{x^*} \neq 0$, the functional equilibrium of $x^* = a_{i_1}^{(s)} \in S_{i_1}$ with $g(x^*) = a_{i_1}$ is unstable (a functional *lower-saddle* of the fourth-order, $d^4 f/dx^4|_{x^*} < 0$). Such a flow is called a functional *lower-saddle* flow of the fourth-order. The bifurcation of equilibrium at $x^* = a_{i_1}^{(s)}$ for two simple functional equilibriums switching to four simple functional equilibriums is called a fourth-order functional *lower-saddle-node* at a point $\mathbf{p} = \mathbf{p}_1$.

From the previous definition, the stability and bifurcations of equilibriums in the 1-dimensional, quartic nonlinear functional dynamical system ($a_0 > 0$ and $dg/dx|_{x^*} > 0$) is presented in Fig. 4.11. In Figs. 4.11a–c, the functional *upper-saddle-node* (USN) and functional *lower-saddle-node* (LSN) switching bifurcations are at two locations for two simple functional

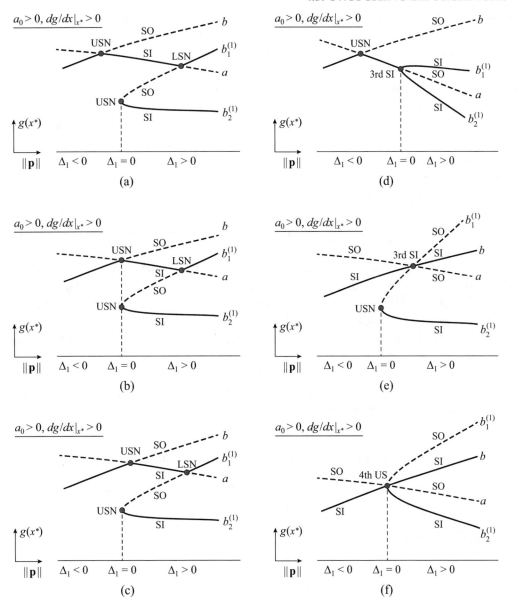

Figure 4.11: Stability and bifurcations of functional equilibriums in the 1-dimensional, quartic nonlinear functional dynamical system ($a_0 > 0$ and $dg/dx|_{x*} > 0$): (a)–(c) Two switching and one appearing bifurcations, (d) 3rd SI switching plus US appearing bifurcations, (e) 3rd SO switching plus US appearing bifurcation, (f) 4th US bifurcation. LSN: *lower-saddle-node*, USN: *upper-saddle-node*, SI: sink, SO: source. Stable and unstable equilibriums are represented by solid and dashed curves, respectively. The bifurcation points are marked by circular symbols.

equilibriums, and one functional *upper-saddle-node* (USN) appearing bifurcation is for two simple functional equilibriums. In Fig. 4.11d, a third-order functional sink (3rd SI) pitchfork-switching bifurcation for a switching of one functional sink equilibrium to three simple functional equilibriums is presented, and one functional *upper-saddle-node* (USN) switching bifurcation for two simple functional equilibriums switching is also presented. In Fig. 4.11e, a third-order functional source (3rd SO) bundle-switching bifurcation for three simple equilibrium bundle-switching is presented, and a functional *upper-saddle-node* (USN) appearing bifurcation for two functional equilibrium onset is also presented. In Fig. 4.11f, a fourth-order functional *upper-saddle* (4th US) flower-bundle switching bifurcation for four simple functional equilibriums are presented.

Similarly, the stability and bifurcations of equilibriums in the 1-dimensional, quartic nonlinear dynamical system ($a_0 < 0$ and $dg/dx|_{x^*} > 0$) is presented in Fig. 4.12. In Figs. 4.12a–c, functional *lower-saddle-node* (LSN) and functional *upper-saddle-node* (USN) switching bifurcations are at two locations for two simple functional equilibriums, and one functional *lower-saddle-node* (LSN) appearing bifurcation is for two simple functional equilibriums appearing. In Fig. 4.12d, a third-order source (3rd SO) pitchfork-switching bifurcation for a switching of one source equilibrium to three simple equilibriums is presented, and one *lower-saddle-node* (LSN) switching bifurcation for two simple equilibriums switching is also presented. In Fig. 4.12e, a third-order sink (3rd SI) bundle-switching bifurcation for three equilibrium bundle-switching is presented, and a functional *lower-saddle-node* (LSN) appearing bifurcation for two functional equilibrium onset is also presented. In Fig. 4.12f, a fourth-order functional *lower-saddle* (4th LS) flower-bundle switching bifurcation for four simple functional equilibriums are presented.

For the further discussion on the switching bifurcation, the following definition is given for the 1-dimensional, quartic nonlinear dynamical system.

Definition 4.6   Consider a 1-dimensional, quartic nonlinear functional dynamical system

$$\dot{x} = A(\mathbf{p})(g(x))^4 + B(\mathbf{p})(g(x))^3 + C(\mathbf{p})(g(x))^2 + D(\mathbf{p})g(x) + E(\mathbf{p})$$
$$= a_0(\mathbf{p})(g(x) - a)(g(x) - b)(g(x) - c)(g(x) - d), \tag{4.116}$$

where $A(\mathbf{p}) \neq 0$, and

$$\mathbf{p} = (p_1, p_2, \ldots, p_m)^{\mathrm{T}}. \tag{4.117}$$

(i) If

$$S_i = \{a_i^{(s)} | g(a_i^{(s)}) = a_i, \ s = 1, 2, \ldots, N_i\} \cup \{\emptyset\}, \ i = 1, 2, 3, 4;$$
$$\{a_1, a_2, a_3, a_4\} = \mathrm{sort}\{a, b, c, d\}, \ a_i \leq a_{i+1}, \tag{4.118}$$

the quartic nonlinear functional dynamical system has any four simple functional equilibriums. The standard functional form is

$$\dot{x} = a_0(\mathbf{p})(g(x) - a_1)(g(x) - a_2)(g(x) - a_3)(g(x) - a_4). \tag{4.119}$$

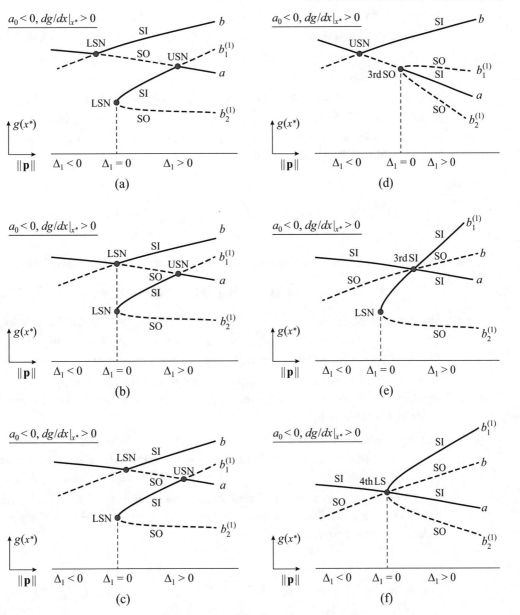

Figure 4.12: Stability and bifurcations of functional equilibriums in the 1-dimensional, quartic nonlinear functional dynamical system ($a_0 < 0$ and $dg/dx|_{x^*} > 0$): (a)–(c) Two switching and one appearing bifurcations, (d) 3rd SI switching plus US appearing bifurcations, (e) 3rd SO switching plus US appearing bifurcation, (f) 4th US bifurcation. LSN: *lower-saddle-node*, USN: *upper-saddle-node*, SI: sink, SO: source. Stable and unstable equilibriums are represented by solid and dashed curves, respectively. The bifurcation points are marked by circular symbols.

(i_1) For $a_0 dg/dx|_{x*} > 0$, the equilibrium flow is a functional (SI:SO:SI:SO) flow. The functional equilibrium of $x^* = a_{1,3}^{(s)} \in S_{1,3}$ with $g(x^*) = a_{1,3}$ is stable (sink, $df/dx|_{x*} < 0$) and the equilibrium of $x^* = a_{2,4}^{(s)} \in S_{2,4}$ with $g(x^*) = a_{2,4}$ is unstable (source, $df/dx|_{x*} > 0$).

(i_2) For $a_0 dg/dx|_{x*} < 0$, the equilibrium flow is a functional (SO:SI:SO:SI) flow. The functional equilibrium of $x^* = a_{1,3}^{(s)} \in S_{1,3}$ with $g(x^*) = a_{1,3}$ is unstable (source, $df/dx|_{x*} > 0$) and the functional equilibrium of $x^* = a_{2,4}^{(s)} \in S_{2,4}$ with $g(x^*) = a_{2,4}$ is stable (sink, $df/dx|_{x*} < 0$).

(ii) If

$$S_i = \{a_i^{(s)}|g(a_i^{(s)}) = a_i, \ s = 1, 2, \ldots, N_i\} \cup \{\emptyset\}, \ i = \{i_1, i_2, i_3, i_4\};$$
$$\Delta_{i_1 i_2} = (a_{i_1} - a_{i_2})^2 = 0 \text{ with } a_{i_1} = a_{i_2}; i_1, i_2 \in \{1, 2, 3, 4\}$$
(4.120)

the quartic nonlinear functional dynamical system has a standard functional form as

$$\dot{x} = f(x, \mathbf{p}) = a_0 (g(x) - a_{i_1})^2 (g(x) - a_{i_3})(g(x) - a_{i_4}).$$
(4.121)

(ii_1) The equilibrium of $x^* = a_{i_1}^{(s)} \in S_{i_1}$ with $g(x^*) = a_{i_1}$ is unstable (a functional *upper-saddle* of second order, $d^2 f/dx^2|_{x*} > 0$). Such a flow is called a functional *upper-saddle* flow at $x^* = a_{i_1}^{(s)}$. The bifurcation of functional equilibrium at $x^* = a_{i_1}^{(s)}$ for two functional equilibriums switching of $x^* = a_{i_1}^{(s)}, a_{i_2}^{(s)}$ with $g(x^*) = a_{i_1}, a_{i_2}$ is called a functional *upper-saddle-node* bifurcation of the second order at a point $\mathbf{p} = \mathbf{p}_1$.

(ii_2) The equilibrium of $x^* = a_{i_1}^{(s)} \in S_{i_1}$ with $g(x^*) = a_{i_1}$ is unstable (a functional *lower-saddle* of second order, $d^2 f/dx^2|_{x*} < 0$). Such a flow is called a functional *lower-saddle* flow at $x^* = a_{i_1}^{(s)}$. The bifurcation of functional equilibrium at $x^* = a_{i_1}^{(s)}$ for two functional equilibriums switching of $x^* = a_{i_1}^{(s)}, a_{i_2}^{(s)}$ with $g(x^*) = a_{i_1}, a_{i_2}$ is called a functional *lower-saddle-node* bifurcation of the second order at a point $\mathbf{p} = \mathbf{p}_1$.

(iii) If

$$S_i = \{a_i^{(s)}|g(a_i^{(s)}) = a_i, \ s = 1, 2, \ldots, N_i\} \cup \{\emptyset\}, \ i = \{i_1, i_2, i_3, i_4\};$$
$$\Delta_{i_1 i_2} = (a_{i_1} - a_{i_2})^2 = 0, \ \Delta_{i_2 i_3} = (a_{i_2} - a_{i_3})^2 = 0$$
$$a_{i_1} = a_{i_2} = a_3, \ i_1, i_2, i_3 \in \{1, 2, 3, 4\}, \ i_1 \neq i_2 \neq i_3.$$
(4.122)

The corresponding standard functional form is

$$\dot{x} = f(x, \mathbf{p}) = a_0 (g(x) - a_{i_1})^3 (g(x) - a_{i_4})$$
$$i_\alpha \in \{1, 2, 3, 4\}, \ \alpha = 1, 4.$$
(4.123)

(iii$_1$) The functional equilibrium of $x^* = a_{i_1}^{(s)} \in S_{i_1}$ with $g(x^*) = a_{i_1}$ is unstable (a functional source of the third-order, $d^3 f/dx^3|_{x^*} > 0$). Such a flow is called a third-order functional source flow at $x^* = a_{i_1}^{(s)}$. The bifurcation of functional equilibrium at $x^* = a_{i_1}^{(s)}$ for a bundle switching of three functional simple equilibriums of $x^* = a_{i_1}^{(s)}, a_{i_2}^{(s)}, a_{i_3}^{(s)}$ with $g(x^*) = a_{i_1}, a_{i_2}, a_{i_3}$ is called a functional source bifurcation of the third order at a point $\mathbf{p} = \mathbf{p}_1$.

(iii$_2$) The functional equilibrium of $x^* = a_i^{(s)} \in S_{i_1}$ with $g(x^*) = a_{i_1}$ is stable (a functional sink of the third-order, $d^3 f/dx^3|_{x^*} < 0$). Such a flow is called a third-order sink flow at $x^* = a_{i_1}^{(s)}$. The bifurcation of functional equilibrium at $x^* = a_{i_1}^{(s)}$ for a bundle switching of three simple functional equilibriums of $x^* = a_{i_1}^{(s)}, a_{i_2}^{(s)}, a_{i_3}^{(s)}$ with $g(x^*) = a_{i_1}, a_{i_2}, a_{i_3}$ is called a sink bifurcation of the third order at a point $\mathbf{p} = \mathbf{p}_1$.

(iv) If

$$S_i = \{a_i^{(s)} | g(a_i^{(s)}) = a_i, \ s = 1, 2, \ldots, N_i\} \cup \{\emptyset\}, \ i = \{i_1, i_2, i_3, i_4\};$$
$$\Delta_{i_1 i_2} = (a_{i_1} - a_{i_2})^2 = 0, \ \Delta_{i_2 i_3} = (a_{i_2} - a_{i_3})^2 = 0,$$
$$\Delta_{i_3 i_4} = (a_{i_4} - a_{i_4})^2 = 0, \ a_{i_1} = a_{i_2} = a_{i_3} = a_{i_4}, \tag{4.124}$$
$$i_1, i_2, i_3, i_4 \in \{1, 2, 3, 4\}, \ i_1 \neq i_2 \neq i_3 \neq i_4$$

the corresponding standard functional form is

$$\dot{x} = f(x, \mathbf{p}) = a_0 (g(x) - a_{i_1})^4. \tag{4.125}$$

(iv$_1$) The functional equilibrium of $x^* = a_{i_1}^{(s)} \in S_{i_1}$ with $g(x^*) = a_{i_1}$ is unstable (a functional *upper-saddle* of the fourth order, $d^4 f/dx^4|_{x^*} > 0$). Such a flow is called a fourth-order functional *upper-saddle* flow at $x^* = a_{i_1}^{(s)}$. The bifurcation of functional equilibrium at $x^* = a_{i_1}^{(s)}$ for a bundle switching of four simple functional equilibriums of $x^* = a_{i_1}^{(s)}, a_{i_2}^{(s)}, a_{i_3}^{(s)}, a_{i_4}^{(s)}$ with $g(x^*) = a_{1,2,3,4}$ is called a functional *upper-saddle-node* bifurcation of the fourth order at a point $\mathbf{p} = \mathbf{p}_1$.

(iv$_2$) The functional equilibrium of $x^* = a_{i_1}^{(s)} \in S_{i_1}$ with $g(x^*) = a_{i_1}$ is unstable (a functional *lower-saddle* of the fourth order, $d^4 f/dx^4|_{x^*} < 0$). Such a flow is called a fourth-order functional *lower-saddle* flow at $x^* = a_{i_1}^{(s)}$. The bifurcation of functional equilibrium at $x^* = a_{i_1}^{(s)}$ for a bundle switching of four simple functional equilibriums of $x^* = a_{i_1}^{(s)}, a_{i_2}^{(s)}, a_{i_3}^{(s)}, a_{i_4}^{(s)}$ with $g(x^*) = a_{1,2,3,4}$ is called a functional *lower-saddle* bifurcation of the fourth order at a point $\mathbf{p} = \mathbf{p}_1$.

From the previous definition, stability and bifurcations of functional equilibriums in the 1-dimensional, quartic nonlinear functional dynamical system is presented in Fig. 4.13. For $a_0 > 0$

and $dg/dx|_{x*} > 0$, the bifurcations and stability of equilibriums are presented in Figs. 4.13a–c. In Fig. 4.13a, the four functional *upper-saddle-node* (USN) and two functional *lower-saddle-node* (LSN) switching bifurcation network are presented for all possible switching bifurcation between two simple functional equilibriums. In Fig. 4.13b, a third-order functional sink (SI) bundle-switching bifurcation for three simple functional equilibrium is presented, and there are three possible functional *upper-saddle-node* (USN) and functional *lower-saddle-node* (LSN) switching bifurcations for two functional simple equilibriums. In Fig. 4.13c, a fourth-order functional *upper-saddle* (4th US) bundle-switching bifurcation for four simple functional equilibriums are presented. Similarly, for $a_0 < 0$ and $dg/dx|_{x*} > 0$, the bifurcations and stability of functional equilibriums are presented in Figs. 4.13d–f. In Fig. 4.13d, four functional *lower-saddle-node* (LSN) and two functional *upper-saddle-node* (USN) switching bifurcation network are presented for all possible switching bifurcation between two simple functional equilibriums. In Fig. 4.13e, a third-order functional source (SO) bundle-switching bifurcation for three simple functional equilibrium is presented, and there are three possible functional *lower-saddle-node* (LSN) and functional *upper-saddle-node* (USN) switching bifurcations for two simple functional equilibriums. In Fig. 4.13f, a fourth-order *lower-saddle* (4th LS) bundle-switching bifurcation for four simple functional equilibriums are presented.

For the switching bifurcation between the second-order and simple functional equilibriums, the following definition is given for the 1-dimensional, quartic nonlinear functional system.

Definition 4.7   Consider a 1-dimensional, quartic nonlinear functional dynamical system

$$\dot{x} = A(\mathbf{p})(g(x))^4 + B(\mathbf{p})(g(x))^3 + C(\mathbf{p})(g(x))^2 + D(\mathbf{p})g(x) + E(\mathbf{p})$$
$$= a_0(\mathbf{p})(g(x) - a)^2(g(x) - b)(g(x) - c), \tag{4.126}$$

where $A(\mathbf{p}) \neq 0$, and

$$\mathbf{p} = (p_1, p_2, \ldots, p_m)^{\mathrm{T}}. \tag{4.127}$$

(i) If

$$S_i = \{a_i^{(s)} | g(a_i^{(s)}) = a_i, \ s = 1, 2, \ldots, N_i\} \cup \{\emptyset\}, \ i = \{i_1, i_2, i_3\};$$
$$\{a_1, a_2, a_3\} = \text{sort}\{a, b, c\}, \ a_i < a_{i+1}; \ i_1, i_2, i_3 \in \{1, 2, 3\}, \tag{4.128}$$

the quartic nonlinear functional dynamical system has a standard functional form as

$$\dot{x} = f(x, \mathbf{p}) = a_0(g(x) - a_{i_1})^2(g(x) - a_{i_2})(g(x) - a_{i_3}). \tag{4.129}$$

(i_1) The functional equilibrium of $x^* = a_{i_1}^{(s)} \in S_{i_1}$ with $g(x^*) = a_{i_1}$ is unstable (a functional *upper-saddle* of second order, $d^2 f/dx^2|_{x*} > 0$). Such a flow is called a functional *upper-saddle* flow at $x^* = a_{i_1}^{(s)}$.

(i_2) The functional equilibrium of $x^* = a_{i_1}^{(s)} \in S_{i_1}$ with $g(x^*) = a_{i_1}$ is unstable (a functional *lower-saddle* of second order, $d^2 f/dx^2|_{x*} < 0$). Such a flow is called a functional *lower-saddle* flow at $x^* = a_{i_1}^{(s)}$.

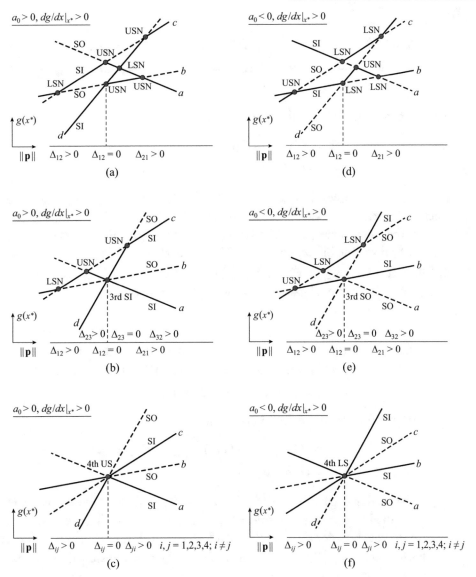

Figure 4.13: Stability and bifurcations of equilibriums in the 1-dimensional, quartic nonlinear dynamical system ($a_0 > 0$ and $dg/dx|_{x*} > 0$): (a) four USN and two LSN switching bifurcation network, (b) third-order SI bundle-switching bifurcation, (c) fourth-order US bundle-switching bifurcation, ($a_0 < 0$ and $dg/dx|_{x*} > 0$): (d) four LSN and two USN switching bifurcation network, (e) third-order SO bundle-switching bifurcation, (f) fourth-order LS bundle-switching bifurcation. LSN: *lower-saddle-node*, USN: *upper-saddle-node*, SI: sink, SO: source. Stable and unstable equilibriums are represented by solid and dashed curves, respectively. The bifurcation points are marked by circular symbols.

(ii) If

$$S_i = \{a_i^{(s)} | g(a_i^{(s)}) = a_i, \ s = 1, 2, \dots, N_i\} \cup \{\emptyset\}, \ i = \{i_1, i_2, i_3\};$$
$$a = a_{i_1}, \ b = a_{i_2}, \ c = a_{i_3}; \ \Delta_{i_2 i_3} = (a_{i_2} - a_{i_3})^2 = 0, \ a_{i_2} = a_{i_3}; \quad (4.130)$$
$$i_1, i_2, i_3 \in \{1, 2\}, \ i_1 \neq i_2 \neq i_3,$$

the corresponding standard functional form is

$$\dot{x} = f(x, \mathbf{p}) = a_0 (g(x) - a_{i_1})^2 (g(x) - a_{i_2})^2$$
$$i_\alpha \in \{1, 2\}, \ \alpha = 1, 3. \quad (4.131)$$

The functional equilibriums of $x^* = a_{i_1}^{(s)}, a_{i_2}^{(s)}$ with $g(x^*) = a_{i_1}, a_{i_2}$ are unstable (a functional *upper-saddle* of the second order, $d^2 f / dx^2|_{x^*} > 0$ or a functional *lower-saddle* of the second order, $d^2 f / dx^2|_{x^*} < 0$). Such a flow is called a functional (US:US) or (LS:LS) flow. The bifurcation of equilibrium at $x^* = a_{i_2}^{(s)}$ for two simple functional equilibriums switching of $x^* = a_{i_1}^{(s)}, a_{i_2}^{(s)}$ with $g(x^*) = a_{i_2}, a_{i_3}$ is called a functional *upper-saddle* or functional *lower-saddle* bifurcation at a point $\mathbf{p} = \mathbf{p}_1$.

(iii) If

$$S_i = \{a_i^{(s)} | g(a_i^{(s)}) = a_i, \ s = 1, 2, \dots, N_i\} \cup \{\emptyset\}, \ i = \{i_1, i_2, i_3\};$$
$$a = a_{i_1}; \ a_{i_3}, a_{i_2} \in \{b, c\}; \ \Delta_{i_1 i_3} = (a_{i_1} - a_{i_3})^2 = 0; \ a_{i_1} = a_{i_3}, \quad (4.132)$$
$$i_1, i_2, i_3 \in \{1, 2, 3\}, \ i_1 \neq i_2 \neq i_3.$$

The corresponding standard functional form is

$$\dot{x} = f(x, \mathbf{p}) = a_0 (g(x) - a_{i_1})^3 (g(x) - a_{i_2})$$
$$i_\alpha \in \{1, 2\}, \ \alpha = 1, 2. \quad (4.133)$$

(iii$_1$) The functional equilibrium of $x^* = a_{i_1}^{(s)} \in S_{i_1}$ with $g(x^*) = a_{i_1}$ is unstable (a functional source of the third-order, $d^3 f / dx^3|_{x^*} > 0$). Such a flow is called a third-order functional source flow at $x^* = a_{i_1}$. The bifurcation of functional equilibrium at $x^* = a_{i_1}^{(s)}$ for a switching of second order and simple functional equilibriums of $x^* = a_{i_2}^{(s)}, a_{i_3}^{(s)}$ is called a functional source switching bifurcation of the third-order at a point $\mathbf{p} = \mathbf{p}_1$.

(iii$_2$) The functional equilibrium of $x^* = a_{i_1}^{(s)} \in S_{i_1}$ with $g(x^*) = a_{i_1}$ is stable (a functional sink of the third order, $d^3 f / dx^3|_{x^*} < 0$). Such a flow is called a third-order functional sink flow at $x^* = a_{i_1}$. The bifurcation of equilibrium at $x^* = a_{i_1}$ for a switching of one second-order and one simple functional equilibriums of $x^* = a_{i_2}^{(s)}, a_{i_3}^{(s)}$ is called a functional sink switching bifurcation of the third order at a point $\mathbf{p} = \mathbf{p}_1$.

(iv) If

$$S_i = \{a_i^{(s)}|g(a_i^{(s)}) = a_i, \ s = 1, 2, \ldots, N_i\} \cup \{\emptyset\}, \ i = \{i_1, i_2, i_3\};$$
$$\Delta_{i_1 i_2} = (a_{i_1} - a_{i_2})^2 = 0; \Delta_{i_2 i_3} = (a_{i_2} - a_{i_3})^2 = 0, \tag{4.134}$$
$$i_1, i_2, i_3 \in \{1, 2, 3\}, \ i_1 \neq i_2 \neq i_3$$

the corresponding standard functional form is

$$\dot{x} = f(x, \mathbf{p}) = a_0(g(x) - a_{i_1})^4. \tag{4.135}$$

(iv$_1$) The functional equilibrium of $x^* = a_{i_1}^{(s)} \in S_{i_1}$ with $g(x^*) = a_{i_1}$ is unstable (a functional *upper-saddle* of the fourth order, $d^4 f/dx^4|_{x^*} > 0$). Such a flow is called a fourth-order *upper-saddle* flow at $x^* = a_{i_1}^{(s)}$. The bifurcation of equilibrium at $x^* = a_{i_1}^{(s)}$ for a bundle switching of one second-order and two simple functional equilibriums of $x^* = a_{i_1}^{(s)}, a_{i_2}^{(s)}, a_{i_3}^{(s)}$ is called a functional *upper-saddle-node* bifurcation of the fourth order at a point $\mathbf{p} = \mathbf{p}_1$.

(iv$_2$) The equilibrium of $x^* = a_{i_1}^{(s)} \in S_{i_1}$ with $g(x^*) = a_{i_1}$ is stable (a sink of the third order, $d^4 f/dx^4|_{x^*} < 0$). Such a flow is called a fourth-order functional *lower-saddle* flow at $x^* = a_{i_1}^{(s)}$. The bifurcation of equilibrium at $x^* = a_{i_1}^{(s)}$ for a bundle switching of one second-order and two simple functional equilibriums of $x^* = a_{i_1}^{(s)}, a_{i_2}^{(s)}, a_{i_3}^{(s)}$ is called a functional *lower-saddle* bifurcation of the fourth-order at a point $\mathbf{p} = \mathbf{p}_1$.

From Definition 4.7, stability and bifurcations of functional equilibriums in the 1-dimensional, quartic nonlinear functional dynamical system is presented in Fig. 4.14. For $a_0 > 0$ and $dg/dx|_{x^*} > 0$, the bifurcations and stability of functional equilibriums are presented in Figs. 4.14a–c. In Fig. 4.14a, there is a switching bifurcation network with two third-order functional source switching bifurcations and one functional *upper-saddle-node* bifurcation. The two third-order functional source (3rd SO) switching bifurcations are for (LS:SO) switching to (US:SO) functional equilibriums and for (SO:US) switching to (SO:LS)-functional equilibriums. The functional *upper-saddle-node* (USN) switching bifurcation is for two simple functional equilibriums. In Fig. 4.14b, a fourth-order functional *upper-saddle* (4th US) bundle-switching bifurcation for (SI:LS:SO) switching to (SO:LS:SI) functional equilibriums. In Fig. 4.14c, a fourth-order functional *upper-saddle* (4th US) bundle-switching bifurcation for (SI:SO:US) switching to (SO:SI:US) functional equilibriums. Similarly, for $a_0 dg/dx|_{x^*} < 0$, the bifurcations and stability of functional equilibriums are presented in Figs. 4.14d–f. In Fig. 4.14d, the switching bifurcation network consists of two third-order functional sink switching bifurcations and one functional *lower-saddle-node* bifurcation. The two third-order functional sink (3rd SI) switching bifurcations are for the (LS:SO) switching to (US:SO)-functional equilibriums and

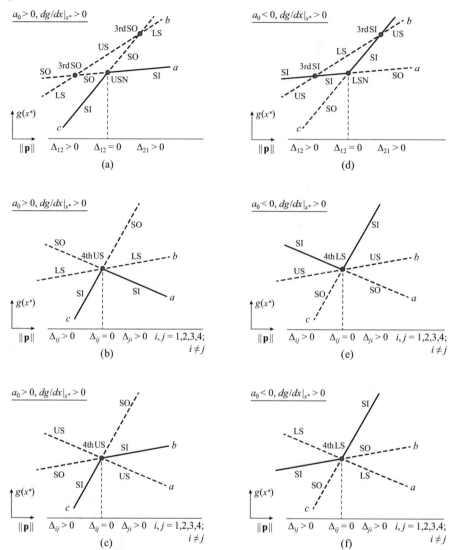

Figure 4.14: Stability and bifurcations of functional equilibriums in the 1-dimensional, quartic nonlinear dynamical system ($a_0 > 0$ and $a_0 dg/dx|_{x*} > 0$): (a) two third-order functional SO and USN switching bifurcation network, (b) a fourth-order functional US bundle switching bifurcation, (c) a fourth-order functional US bundle-switching bifurcation, ($a_0 < 0$ and $dg/dx|_{x*} > 0$): (d) two third-order functional SI and LSN switching bifurcation network, (e) a fourth-order functional LS bundle-switching bifurcation, (f) a fourth-order functional LS bundle switching. LSN: *lower-saddle-node*, USN: *upper-saddle-node*, SI: sink, SO: source. Stable and unstable equilibriums are represented by solid and dashed curves, respectively. The bifurcation points are marked by circular symbols.

for the (SO:US) switching to (SO:LS)-functional equilibriums. The functional *upper-saddle-node* (USN) switching bifurcation is for two simple functional equilibriums. In Fig. 4.14e, a fourth-order *lower-saddle* (4th LS) bundle-switching bifurcation for (SO:US:SI) switching to (SI:US:SO) functional equilibriums. In Fig. 4.14f, a fourth-order *lower-saddle* (4th LS) bundle-switching bifurcation for (SO:SI:LS) switching to (SI:SO:LS) functional equilibriums.

For the switching bifurcation between the third-order and simple functional equilibriums, the following definition is given for the 1-dimensional, quartic nonlinear functional dynamical system.

Definition 4.8   Consider a 1-dimensional, quartic nonlinear functional dynamical system

$$
\begin{aligned}
\dot{x} &= A(\mathbf{p})(g(x))^4 + B(\mathbf{p})(g(x))^3 + C(\mathbf{p})(g(x))^2 + D(\mathbf{p})g(x) + E(\mathbf{p}) \\
&= a_0(\mathbf{p})(g(x) - a)^3(g(x) - b),
\end{aligned}
\tag{4.136}
$$

where $A(\mathbf{p}) \neq 0$, and

$$
\mathbf{p} = (p_1, p_2, \dots, p_m)^{\mathrm{T}}.
\tag{4.137}
$$

(i) If

$$
\begin{aligned}
&S_i = \{a_i^{(s)} | g(a_i^{(s)}) = a_i, \ s = 1, 2, \dots, N_i\} \cup \{\emptyset\}, \ i = \{i_1, i_2\}; \\
&\{a_1, a_2\} = \text{sort}\{a, b\}, \ a_i < a_{i+1}; \ i_1, i_2 \in \{1, 2\},
\end{aligned}
\tag{4.138}
$$

the quartic nonlinear dynamical system has a standard form as

$$
\dot{x} = f(x, \mathbf{p}) = a_0(g(x) - a_{i_1})^3(g(x) - a_{i_2}).
\tag{4.139}
$$

(i$_1$) The functional equilibrium of $x^* = a_{i_1}^{(s)} \in S_{i_1}$ with $g(x^*) = a_{i_1}$ is unstable (a third-order functional source, $d^3 f/dx^3|_{x^*} > 0$). The functional equilibrium of $x^* = a_{i_2}^{(s)} \in S_{i_2}$ with $g(x^*) = a_{i_2}$ is stable (a functional sink, $df/dx|_{x^*} < 0$). Such a flow is called a (3rd SO:SI) or (SI:3rd SO)-flow.

(i$_2$) The functional equilibrium of $x^* = a_{i_1}^{(s)} \in S_{i_1}$ with $g(x^*) = a_{i_1}$ is stable (a third-order functional sink, $d^3 f/dx^3|_{x^*} < 0$). The functional equilibrium of $x^* = a_{i_2}^{(s)} \in S_{i_2}$ with $g(x^*) = a_{i_2}$ is unstable (a functional source, $df/dx|_{x^*=a_{i_2}} > 0$). Such a flow is called a (3rdSI:SO) or (SO:3rd SI)-flow.

(ii) If

$$
\begin{aligned}
&S_i = \{a_i^{(s)} | g(a_i^{(s)}) = a_i, \ s = 1, 2, \dots, N_i\} \cup \{\emptyset\}, \ i = \{i_1, i_2\}; \\
&a = a_{i_1}, \ b = a_{i_2}; \ \Delta_{i_1 i_2} = (a_{i_1} - a_{i_2})^2 = 0, \ a_{i_1} = a_{i_2}; \\
&i_1, i_2 \in \{1, 2\}, \ i_1 \neq i_2,
\end{aligned}
\tag{4.140}
$$

the corresponding standard form is

$$
\dot{x} = f(x, \mathbf{p}) = a_0(g(x) - a_{i_1})^4.
\tag{4.141}
$$

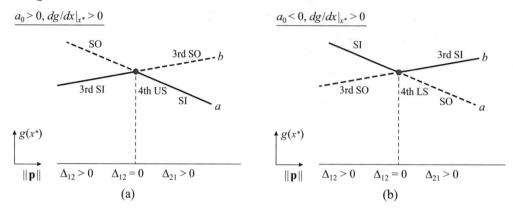

Figure 4.15: Stability and bifurcations of functional equilibriums in the 1-dimensional, quartic nonlinear functional dynamical system ($a_0 > 0, dg/dx|_{x*} > 0$): (a) a fourth-order functional US switching bifurcation of (3rd SI: SO) to (SI: 3rd SO), ($a_0 < 0, dg/dx|_{x*} > 0$): (b) a fourth-order LS switching bifurcation of (3rd SO: SI) to (SO: 3rd SI). LS: *lower-saddle*, US: *upper-saddle*, SI: sink, SO: source. Stable and unstable equilibriums are represented by solid and dashed curves, respectively. The bifurcation points are marked by circular symbols.

(ii$_1$) The functional equilibrium of $x^* = a_{i_1}^{(s)} \in S_{i_1}$ with $g(x^*) = a_{i_1}$ is unstable (a fourth-order *upper-saddle*, $d^4 f/dx^4|_{x*} > 0$). Such a flow is called a fourth-order functional *upper-saddle* flow. The bifurcation of functional equilibrium at $x^* = a_{i_1}^{(s)}$ for a switching of one third-order and one simple functional equilibrium of $x^* = a_1^{(s)}, a_2^{(s)}$ is called a fourth-order functional *upper-saddle* switching bifurcation at a point $\mathbf{p} = \mathbf{p}_1$.

(ii$_2$) The functional equilibrium of $x^* = a_{i_1}^{(s)} \in S_{i_1}$ with $g(x^*) = a_{i_1}$ is unstable (a fourth-order functional *lower-saddle*, $d^4 f/dx^4|_{x*=a_{i_1}} < 0$). Such a flow is called a fourth-order functional *lower-saddle* flow. The bifurcation of functional equilibrium at $x^* = a_{i_1}^{(s)}$ for a switching of one third-order and one simple functional equilibrium of $x^* = a_1^{(s)}, a_2^{(s)}$ is called a fourth-order functional *lower-saddle* switching bifurcation at a point $\mathbf{p} = \mathbf{p}_1$.

From Definition 4.8, the stability and bifurcations of functional equilibriums in the 1-dimensional, quartic nonlinear functional dynamical system is presented in Fig. 4.15. In Fig. 4.15a, the fourth-order functional *upper-saddle* (4th US) switching bifurcation for $a_0 > 0$ and $dg/dx|_{x*} > 0$ is presented for the third-order functional sink (3rd SI) with a simple functional source (SO) equilibriums (i.e., (3rd SI: SO)) to the third-order functional source (3rd SO) with a simple functional sink (SI) equilibriums (i.e., (SI: 3rd SO)). Similarly, in Fig. 4.15b, the fourth-order *lower-saddle* (4th LS) switching bifurcation for $a_0 < 0$ and $dg/dx|_{x*} > 0$ is presented for the third-order functional source (3rd SO) with a simple functional sink (SI) equilibriums (i.e., (3rd SO: SI)) to the third-order functional sink (3rd SI) with a simple functional source (SO) equilibriums (i.e., (SO: 3rd SI)).

CHAPTER 5

(2m)th-Degree Polynomial Functional Systems

In this chapter, the global stability and bifurcation of equilibriums in the 1-dimensional, $(2m)$th-degree polynomial functional nonlinear systems are discussed.

Definition 5.1   Consider a $(2m)$th-degree polynomial nonlinear functional dynamical system

$$
\begin{aligned}
\dot{x} &= A_0(\mathbf{p})(g(x))^{2m} + A_1(\mathbf{p})(g(x))^{2m-1} + \\
&\quad \cdots + A_{2m-2}(\mathbf{p})(g(x))^2 + A_{2m-1}g(x) + A_{2m}(\mathbf{p}) \\
&= a_0(\mathbf{p})[(g(x))^2 + B_1(\mathbf{p})g(x) + C_1(\mathbf{p})] \\
&\quad \cdots [(g(x))^2 + B_m(\mathbf{p})g(x) + C_m(\mathbf{p})],
\end{aligned}
\tag{5.1}
$$

where $A_0(\mathbf{p}) \neq 0$ and

$$
\mathbf{p} = (p_1, p_2, \ldots, p_m)^{\mathrm{T}}.
\tag{5.2}
$$

(i) If

$$
\Delta_i = B_i^2 - 4C_i < 0 \text{ for } i = 1, 2, \ldots, m,
\tag{5.3}
$$

the 1-dimensional nonlinear functional dynamical system with a $(2m)$th-degree polynomial does not have any equilibrium, and the corresponding standard functional form is

$$
\dot{x} = a_0[(g(x) + \tfrac{1}{2}B_1)^2 + \tfrac{1}{4}(-\Delta_1)] \ldots [(g(x) + \tfrac{1}{2}B_m)^2 + \tfrac{1}{4}(-\Delta_m)].
\tag{5.4}
$$

The flow of such a functional system without equilibriums is called a non-equilibrium flow.

(a) If $a_0 > 0$, the non-equilibrium flow is called a positive flow.

(b) If $a_0 < 0$, the non-equilibrium flow is called a negative flow.

(ii) If

$$
\begin{aligned}
\Delta_i &= B_i^2 - 4C_i > 0 \quad i = i_1, i_2, \ldots, i_l \in \{1, 2, \ldots, m\}, \\
\Delta_j &= B_j^2 - 4C_j < 0 \quad j = i_{l+1}, i_{l+2}, \ldots, i_m \in \{1, 2, \ldots, m\} \\
&\quad \text{with } l \in \{0, 1, \ldots, m\},
\end{aligned}
\tag{5.5}
$$

the 1-dimensional, $(2m)$th-degree polynomial functional dynamical system has $2l$-functional equilibriums as

$$g(x^*) = b_1^{(i)} = -\frac{1}{2}(B_i + \sqrt{\Delta_i}),$$

$$g(x^*) = b_2^{(i)} = -\frac{1}{2}(B_i - \sqrt{\Delta_i}),$$

$$i \in \{i_1, i_2, \ldots, i_l\} \subseteq \{1, 2, \ldots, m\}, \tag{5.6}$$

$$\{a_1, a_2, \ldots, a_{2l}\} = \text{sort} \{b_1^{(1)}, b_2^{(1)}, \ldots, b_1^{(l)}, b_2^{(l)}\}, \ a_\alpha < a_{\alpha+1},$$

$$S_\alpha = \{a_\alpha^{(s_\alpha)} | g(a_\alpha^{(s_\alpha)}) = a_\alpha, \ s_\alpha = 1, 2, \ldots, N_\alpha\} \cup \{\emptyset\},$$

$$\alpha = \{1, 2, \ldots, 2l\}.$$

(ii$_1$) If

$$b_r^{(i)} \neq b_s^{(j)} \text{ for } r, s \in \{1, 2\}; \ i, j = 1, 2, \ldots, l;$$

$$\{a_1, a_2, \ldots, a_{2l}\} = \text{sort} \{b_1^{(1)}, b_2^{(1)}, \ldots, b_1^{(l)}, b_2^{(l)}\}, \ a_\alpha < a_{\alpha+1}, \tag{5.7}$$

then, the corresponding standard functional form is

$$\dot{x} = a_0 \prod_{i=1}^{l} (g(x) - a_{2i-1})(g(x) - a_{2i}) \prod_{k=l+1}^{m} [(g(x) + \frac{1}{2}B_{i_k})^2 + \frac{1}{4}(-\Delta_{i_k})]. \tag{5.8}$$

(a) If $a_0 dg/dx|_{x^*} > 0$, the simple functional equilibrium separatrix flow is called a functional (SI:SO: …:SI:SO: …:SI:SO)-flow.

(b) If $a_0 dg/dx|_{x^*} < 0$, the simple functional equilibrium separatrix flow is called a (SO:SI: …:SO:SI: …:SO:SI)-flow.

(ii$_2$) If

$$\{a_1, a_2, \ldots, a_{2l}\} = \text{sort} \{b_1^{(1)}, b_2^{(1)}, \ldots, b_1^{(l)}, b_2^{(l)}\},$$

$$a_{i_1} \equiv a_1 = \cdots = a_{l_1},$$

$$a_{i_2} \equiv a_{l_1+1} = \cdots = a_{l_1+l_2},$$

$$\vdots \tag{5.9}$$

$$a_{i_r} \equiv a_{\sum_{i=1}^{r-1} l_i + 1} = \cdots = a_{\sum_{i=1}^{r-1} l_i + l_r} = a_{2l}$$

$$\text{with } \sum_{s=1}^{r} l_s = 2l,$$

then, the corresponding standard functional form is

$$\dot{x} = a_0 \prod_{s=1}^{r} (g(x) - a_{i_s})^{l_s} \prod_{k=l+1}^{m} [(g(x) + \frac{1}{2}B_{i_k})^2 + \frac{1}{4}(-\Delta_{i_k})]. \tag{5.10}$$

The functional equilibrium separatrix flow is called a $(l_1 \text{th XX} : l_2 \text{th XX} : \ldots : l_r \text{th XX})$-flow.

(a_1) For $a_0 dg/dx|_{x^*} > 0$ $(p = 1, 2, \ldots, r)$,

$$l_p \text{th XX} = \begin{cases} (2r_p - 1)^{\text{th}} \text{ order source, for } \alpha_p = 2M_p - 1, \, l_p = 2r_p - 1; \\ (2r_p - 1)^{\text{th}} \text{ order sink, for } \alpha_p = 2M_p, \, l_p = 2r_p - 1, \end{cases}$$

(5.11)

where

$$\alpha_p = \sum_{s=p}^{r} l_s.$$

(5.12)

(a_2) For $a_0 dg/dx|_{x^*} < 0$ $(p = 1, 2, \ldots, r)$,

$$l_p \text{th XX} = \begin{cases} (2r_p - 1)^{\text{th}} \text{ order sink, for } \alpha_p = 2M_p - 1, \, l_p = 2r_p - 1; \\ (2r_p - 1)^{\text{th}} \text{ order source, for } \alpha_p = 2M_p, \, l_p = 2r_p - 1. \end{cases}$$

(5.13)

(a_3) For $a_0 > 0$ and $dg/dx|_{x^*} \neq 0$ $(p = 1, 2, \ldots, r)$,

$$l_p \text{th XX} = \begin{cases} (2r_p)^{\text{th}} \text{ order lower-saddle, for } \alpha_p = 2M_p - 1, \, l_p = 2r_p; \\ (2r_p)^{\text{th}} \text{ order upper-saddle, for } \alpha_p = 2M_p, \, l_p = 2r_p - 1. \end{cases}$$

(5.14)

(a_4) For $a_0 < 0$ and $dg/dx|_{x^*} \neq 0$ $(p = 1, 2, \ldots, r)$,

$$l_p \text{th XX} = \begin{cases} (2r_p)^{\text{th}} \text{ order upper-saddle, for } \alpha_p = 2M_p - 1, \, l_p = 2r_p; \\ (2r_p)^{\text{th}} \text{ order lower-saddle, for } \alpha_p = 2M_p, \, l_p = 2r_p. \end{cases}$$

(5.15)

(b) The functional equilibrium of $x^* = a_{i_p}^{(s)} \in S_{i_p}$ with $g(x^*) = a_{i_p}$ for $(l_p > 1)$-repeated functional equilibriums switching $(p = 1, 2, \ldots, r)$ is called an $l_p \text{th XX}$ functional bifurcation of $(l_{p_1} \text{th XX} : l_{p_2} \text{th XX} : \ldots : l_{p_\beta} \text{th XX})$-functional equilibrium switching at a point $\mathbf{p} = \mathbf{p}_1 \in \partial\Omega_{12}$, and the functional bifurcation condition is

$$a_{i_p} \equiv a_{\sum_{i=1}^{p-1} l_i + 1} = \cdots = a_{\sum_{i=1}^{p-1} l_i + l_p},$$
$$a^{\pm}_{\sum_{i=1}^{p-1} l_i + 1} \neq \cdots \neq a^{\pm}_{\sum_{i=1}^{p-1} l_i + l_p};$$
$$l_p = \sum_{i=1}^{\beta} l_{p_i}, \quad p = 1, 2, \ldots, r.$$

(5.16)

(iii)

$$\Delta_i = B_i^2 - 4C_i = 0, \ i \in \{i_{11}, i_{12}, \ldots, i_{l_s}\} \subseteq \{i_1, i_2, \ldots, i_l\} \subseteq \{1, 2, \ldots, m\},$$
$$\Delta_k = B_k^2 - 4C_k > 0, \ k \in \{i_{21}, i_{22}, \ldots, i_{2r}\} \subseteq \{i_1, i_2, \ldots, i_l\} \subseteq \{1, 2, \ldots, m\}, \quad (5.17)$$
$$\Delta_j = B_j^2 - 4C_j < 0, \ j \in \{i_{l+1}, i_{l+2}, \ldots, i_m\} \subseteq \{1, 2, \ldots, m\},$$

the 1-dimensional, $(2m)$th-degree polynomial functional dynamical system has $2l$-functional equilibriums as

$$S_\alpha = \{a_\alpha^{(s_\alpha)} | g(a_\alpha^{(s_\alpha)}) = a_\alpha, \ s_\alpha = 1, 2, \ldots, N_\alpha\} \cup \{\emptyset\},$$
$$\alpha = \{1, 2, \ldots, 2l\};$$
$$\{a_1, a_2, \ldots, a_{2l}\} = \mathrm{sort}\{b_1^{(1)}, b_2^{(1)}, \ldots, b_1^{(l)}, b_2^{(l)}\},$$
$$g(x^*) = b_1^{(i)} = -\frac{1}{2}B_i, \ g(x^*) = b_2^{(i)} = -\frac{1}{2}B_i, \ \text{for } i \in \{i_{11}, i_{12}, \ldots, i_{ls}\}, \quad (5.18)$$
$$g(x^*) = b_1^{(k)} = -\frac{1}{2}(B_k + \sqrt{\Delta_k}), \ g(x^*) = b_2^{(k)} = -\frac{1}{2}(B_k - \sqrt{\Delta_k})$$
$$\text{for } k \in \{i_{21}, i_{22}, \ldots, i_{2r}\}.$$

If

$$\{a_1, a_2, \ldots, a_{2l}\} = \mathrm{sort}\{b_1^{(1)}, b_2^{(1)}, \ldots, b_1^{(l)}, b_2^{(l)}\},$$
$$a_{i_1} \equiv a_1 = \cdots = a_{l_1},$$
$$a_{i_2} \equiv a_{l_1+1} = \cdots = a_{l_1+l_2},$$
$$\vdots \quad (5.19)$$
$$a_{i_r} \equiv a_{\sum_{i=1}^{r-1} l_i + 1} = \cdots = a_{\sum_{i=1}^{r-1} l_i + l_r} = a_{2l}$$
$$\text{with } \sum_{s=1}^{r} l_s = 2l,$$

then, the corresponding standard form is

$$\dot{x} = a_0 \prod_{s=1}^{r} (x - a_{i_s})^{l_s} \prod_{k=l+1}^{m} [(x + \frac{1}{2}B_{i_k})^2 + \frac{1}{4}(-\Delta_{i_k})]. \quad (5.20)$$

The functional equilibrium separatrix flow is called a $(l_1$th XX : l_2th XX : $\ldots$: l_rth XX)-flow.

(a) The functional equilibrium of $x^* = a_{i_p}^{(s)} \in S_{i_p}$ with $g(x^*) = a_{i_p}$ for $(l_p > 1)$-repeated functional equilibriums appearance or vanishing $(p = 1, 2, \ldots, r)$ is called an l_pth XX bifurcation of functional equilibrium at a point $\mathbf{p} = \mathbf{p}_1 \in \partial\Omega_{12}$, and the functional bifurcation condition is

$$a_{i_p} \equiv a_{\sum_{i=1}^{p-1} l_i + 1} = \cdots = a_{\sum_{i=1}^{p-1} l_i + l_p} = -\frac{1}{2} B_{i_p},$$

$$\text{with } \Delta_{i_p} = B_{i_p}^2 - 4C_{i_p} = 0 \, (i_p \in \{i_1, i_2, \ldots i_l\}), \tag{5.21}$$

$$a^+_{\sum_{i=1}^{p-1} l_i + 1} \neq \cdots \neq a^+_{\sum_{i=1}^{p-1} l_i + l_p} \text{ or } a^-_{\sum_{i=1}^{p-1} l_i + 1} \neq \cdots \neq a^-_{\sum_{i=1}^{p-1} l_i + l_p}.$$

(b) The functional equilibrium of $x^* = a_{i_q}^{(s)} \in S_{i_q}$ with $g(x^*) = a_{i_q}$ for $(l_q > 1)$-repeated functional equilibriums switching $(q = 1, 2, \ldots, r_1)$ is called an l_qth XX functional bifurcation of $(l_{q_1}$th XX $: l_{q_2}$th XX $: \ldots : l_{q_\beta}$th XX$)$ functional equilibrium switching at a point $\mathbf{p} = \mathbf{p}_1 \in \partial\Omega_{12}$, and the functional bifurcation condition is

$$a_{i_q} \equiv a_{\sum_{i=1}^{q-1} l_i + 1} = \cdots = a_{\sum_{i=1}^{q-1} l_i + l_q},$$

$$a^{\pm}_{\sum_{i=1}^{q-1} l_i + 1} \neq \cdots \neq a^{\pm}_{\sum_{i=1}^{q-1} l_i + l_q}; \tag{5.22}$$

$$l_q = \sum_{i=1}^{\beta} l_{q_i}, \, q = 1, 2, \ldots, r_1.$$

(c) The functional equilibrium of $x^* = a_{i_p}^{(s)} \in S_{i_p}$ with $g(x^*) = a_{i_p}$ for $(l_{p_1} \geq 1)$-repeated functional equilibriums appearance/vanishing and $(l_{p_2} \geq 2)$ repeated functional equilibriums switching of $(l_{p_{21}}$th XX $: l_{p_{22}}$th XX $: \ldots : l_{p_{2\beta}}$th XX$)$-functional equilibrium switching is called an l_pth XX bifurcation of functional equilibrium at a point $\mathbf{p} = \mathbf{p}_1 \in \partial\Omega_{12}$, and the functional bifurcation condition is

$$a_{i_p} \equiv a_{\sum_{i=1}^{p-1} l_i + 1} = \cdots = a_{\sum_{i=1}^{p-1} l_i + l_p}$$

$$\text{with } \Delta_{i_q} = B_{i_q}^2 - 4C_{i_q} = 0 \, (i_q \in \{i_1, i_2, \ldots i_l\})$$

$$a^+_{\sum_{i=1}^{p-1} l_i + j_1} \neq \cdots \neq a^+_{\sum_{i=1}^{p-1} l_i + j_{p_1}} \text{ or } a^-_{\sum_{i=1}^{p_1-1} l_i + j_1} \neq \cdots \neq a^-_{\sum_{i=1}^{p_1-1} l_i + j_{p_1}},$$

$$\text{for } \{j_1, j_2, \ldots, j_{p_1}\} \subseteq \{1, 2, \ldots, l_p\},$$

$$a^{\pm}_{\sum_{i=1}^{p-1} l_i + k_1} \neq \cdots \neq a^{\pm}_{\sum_{i=1}^{p-1} l_i + k_{p_2}}$$

$$\text{for } \{k_1, k_2, \ldots, k_{p_2}\} \subseteq \{1, 2, \ldots, l_p\},$$

$$\text{with } l_{p_1} + l_{p_2} = l_p; \, l_{p_2} = \sum_{i=1}^{\beta} l_{p_{2i}}, \, p = 1, 2, \ldots, r.$$

$$\tag{5.23}$$

(iv) If

$$\Delta_i = B_i^2 - 4C_i > 0 \text{ for } i = 1, 2, \ldots, m \tag{5.24}$$

the 1-dimensional, $(2m)$th-degree polynomial functional dynamical system has $(2m)$ functional equilibriums as

$$S_\alpha = \{a_\alpha^{(s_\alpha)} | g(a_\alpha^{(s_\alpha)}) = a_\alpha, \ s_\alpha = 1, 2, \ldots, N_\alpha\} \cup \{\emptyset\},$$

$$\alpha = \{1, 2, \ldots, 2m\};$$

$$\{a_1, a_2, \ldots, a_{2m}\} = \text{sort}\{b_1^{(1)}, b_2^{(1)}, \ldots, b_1^{(m)}, b_2^{(m)}\}, \ a_s < a_{s+1}$$

$$g(x^*) = b_1^{(i)} = -\frac{1}{2}(B_i + \sqrt{\Delta_i}),$$

$$g(x^*) = b_2^{(i)} = -\frac{1}{2}(B_i - \sqrt{\Delta_i})$$

(5.25)

for $i = 1, 2, \ldots, m$.

(iv$_1$) If

$$b_r^{(i)} \neq b_s^{(j)} \text{ for } r, s \in \{1, 2\}; \ i, j = 1, 2, \ldots, m$$

$$\{a_1, a_2, \ldots, a_{2m}\} = \text{sort}\{b_1^{(1)}, b_2^{(1)}, \ldots, b_1^{(m)}, b_2^{(m)}\}, \ a_s < a_{s+1}.$$

(5.26)

The corresponding standard form is

$$\dot{x} = a_0(g(x) - a_1)(g(x) - a_2)(g(x)a_3)(g(x) - a_4)$$
$$\cdots (g(x) - a_{2m-1})(g(x) - a_{2m}).$$

(5.27)

Such a flow is formed with all the simple equilibriums.

(a) If $a_0 dg/dx|_{x^*} > 0$, the simple functional equilibrium separatrix flow is called a functional (SO:SI: ...: SO:SI: ...:SO:SI)-flow.

(b) If $a_0 dg/dx|_{x^*} < 0$, the simple functional equilibrium separatrix flow is called a functional (SI:SO: ...: SI:SO: ...:SI:SO)-flow.

(iv$_2$) If

$$\{a_1, a_2, \ldots, a_{2m}\} = \text{sort}\{b_1^{(1)}, b_2^{(1)}, \ldots, b_1^{(m)}, b_2^{(m)}\},$$

$$a_{i_1} \equiv a_1 = \cdots = a_{l_1},$$

$$a_{i_2} \equiv a_{l_1+1} = \cdots = a_{l_1+l_2},$$

$$\vdots$$

$$a_{i_r} \equiv a_{\sum_{i=1}^{r-1} l_i + 1} = \cdots = a_{\sum_{i=1}^{r-1} l_i + l_r} = a_{2m}$$

$$\text{with } \sum_{s=1}^{r} l_s = 2m,$$

(5.28)

then, the corresponding standard functional form is

$$\dot{x} = a_0 \prod_{s=1}^{r} (g(x) - a_{i_s})^{l_s}. \tag{5.29}$$

The functional equilibrium separatrix flow is called an $(l_1 \text{th XX} : l_2 \text{th XX} : \ldots : l_r \text{th XX})$-flow. The functional equilibrium of $x^* = a_{i_p}^{(s)} \in S_{i_p}$ with $g(x^*) = a_{i_p}$ for l_p-repeated functional equilibriums switching $(p = 1, 2, \ldots, r)$ is called an $l_p \text{th XX}$ functional bifurcation of $(l_{p_1} \text{th XX} : l_{p_2} \text{th XX} : \ldots : l_{p_\beta} \text{th XX})$ functional equilibrium switching at a point $\mathbf{p} = \mathbf{p}_1 \in \partial\Omega_{12}$, and the functional bifurcation condition is

$$a_{i_p} \equiv a_{\sum_{i=1}^{p-1} l_i + 1} = \cdots = a_{\sum_{i=1}^{p-1} l_i + l_p},$$
$$a_{\sum_{i=1}^{p-1} l_i + 1}^{\pm} \neq \cdots \neq a_{\sum_{i=1}^{p-1} l_i + l_p}^{\pm}; \tag{5.30}$$
$$l_p = \sum_{i=1}^{\beta} l_{p_i}, \quad p = 1, 2, \ldots, r.$$

Definition 5.2 Consider a 1-dimensional, $(2m)$th-degree polynomial nonlinear functional dynamical system as

$$\dot{x} = A_0(\mathbf{p})(g(x))^{2m} + A_1(\mathbf{p})(g(x))^{2m-1} +$$
$$\cdots + A_{2m-2}(\mathbf{p})(g(x))^2 + A_{2m-1}g(x) + A_{2m}(\mathbf{p})$$
$$= a_0(\mathbf{p}) \prod_{i=1}^{n} [(g(x))^2 + B_i(\mathbf{p})g(x) + C_i(\mathbf{p})]^{q_i}, \tag{5.31}$$

where $A_0(\mathbf{p}) \neq 0$, and

$$\mathbf{p} = (p_1, p_2, \ldots, p_{m_1})^{\mathrm{T}}, \quad m = \sum_{i=1}^{n} q_i. \tag{5.32}$$

(i) If

$$\Delta_i = B_i^2 - 4C_i < 0 \text{ for } i = 1, 2, \ldots, n \tag{5.33}$$

the 1-dimensional nonlinear functional dynamical system with a $(2m)$th-degree polynomial does not have any equilibrium, and the corresponding standard functional form is

$$\dot{x} = a_0 \prod_{i=1}^{n} [(g(x) + \frac{1}{2}B_i)^2 + \frac{1}{4}(-\Delta_i)]^{q_i}. \tag{5.34}$$

The flow of such a functional system without functional equilibriums is called a non-functional equilibrium flow.

(a) If $a_0 > 0$, the non-functional equilibrium flow is called the positive flow.

(b) If $a_0 < 0$, the non-functional equilibrium flow is called the negative flow.

(ii) If

$$\Delta_i = B_i^2 - 4C_i > 0, \ i \in \{i_1, i_2, \ldots, i_l\} \subseteq \{1, 2, \ldots, n\},$$
$$\Delta_j = B_j^2 - 4C_j < 0, \ j \in \{i_{l+1}, i_{l+2}, \ldots, i_n\} \subseteq \{1, 2, \ldots, n\},$$

(5.35)

the 1-dimensional, $(2m)$th-degree polynomial functional dynamical system has $2l$-functional equilibriums as

$$S_\alpha = \{a_\alpha^{(s_\alpha)} | g(a_\alpha^{(s_\alpha)}) = a_\alpha, \ s_\alpha = 1, 2, \ldots, N_\alpha\} \cup \{\emptyset\},$$
$$\alpha = \{1, 2, \ldots, 2l\};$$
$$\{a_1, a_2, \ldots, a_{2l}\} = \text{sort} \{b_1^{(1)}, b_2^{(1)}, \ldots, b_1^{(l)}, b_2^{(l)}\}, \ a_s < a_{s+1};$$
$$g(x^*) = b_1^{(i)} = -\frac{1}{2}(B_i + \sqrt{\Delta_i}),$$
$$g(x^*) = b_2^{(i)} = -\frac{1}{2}(B_i - \sqrt{\Delta_i})$$
$$i \in \{i_1, i_2, \ldots, i_l\} \subseteq \{1, 2, \ldots, n\}.$$

(5.36)

(ii$_1$) If

$$b_r^{(i)} \neq b_s^{(j)} \text{ for } r, s \in \{1, 2\}; \ i, j = 1, 2, \ldots, l$$
$$\{a_1, a_2, \ldots, a_{2l}\} = \text{sort} \{b_1^{(1)}, b_2^{(1)}, \ldots, b_1^{(l)}, b_2^{(l)}\}, \ a_s < a_{s+1},$$

(5.37)

then, the corresponding standard functional form is

$$\dot{x} = a_0 \prod_{s=1}^{2l} (g(x) - a_s)^{l_s} \prod_{k=l+1}^{n} [(g(x) + \frac{1}{2}B_{i_k})^2 + \frac{1}{4}(-\Delta_{i_k})]^{q_{i_k}}$$

(5.38)

with $l_s \in \{q_{i_1}, q_{i_2}, \ldots, q_{i_l}\}$.

The functional equilibrium separatrix flow is called an $(l_1 \text{th XX} : l_2 \text{th XX} : \ldots : l_{2l} \text{th XX})$-flow.

(a$_1$) For $a_0 dg/dx|_{x^*} > 0 \ (p = 1, 2, \ldots, 2l)$,

$$l_p \text{th XX} = \begin{cases} (2r_p - 1)^{\text{th}} \text{ order source, for } \alpha_p = 2M_p - 1, \ l_p = 2r_p - 1; \\ (2r_p - 1)^{\text{th}} \text{ order sink, for } \alpha_p = 2M_p, \ l_p = 2r_p - 1, \end{cases}$$

(5.39)

where

$$\alpha_p = \sum_{s=p}^{2l} l_s.$$

(5.40)

(a$_2$) For $a_0 dg/dx|_{x*} < 0$ $(p = 1, 2, \ldots, 2l)$,

$$
l_p \text{th XX} = \begin{cases} (2r_p^{-1})^{\text{th}} \text{ order sink, for } \alpha_p = 2M_p - 1, \ l_p = 2r_p - 1; \\ (2r_p^{-1})^{\text{th}} \text{ order source, for } \alpha_p = 2M_p, \ l_p = 2r_p - 1. \end{cases}
$$

(5.41)

(a$_3$) For $a_0 > 0$ and $dg/dx|_{x*} \neq 0$ $(p = 1, 2, \ldots, 2l)$,

$$
l_p \text{th XX} = \begin{cases} (2r_p)^{\text{th}} \text{ order lower-saddle, for } \alpha_p = 2M_p - 1, \ l_p = 2r_p; \\ (2r_p)^{\text{th}} \text{ order upper-saddle, for } \alpha_p = 2M_p, \ l_p = 2r_p. \end{cases}
$$

(5.42)

(a$_4$) For $a_0 < 0$ and $dg/dx|_{x*} \neq 0$ $(p = 1, 2, \ldots, 2l)$,

$$
l_p \text{th XX} = \begin{cases} (2r_p)^{\text{th}} \text{ order upper-saddle, for } \alpha_p = 2M_p - 1, \ l_p = 2r_p; \\ (2r_p)^{\text{th}} \text{ order lower-saddle, for } \alpha_p = 2M_p, \ l_p = 2r_p. \end{cases}
$$

(5.43)

(ii$_2$) If

$$
\{a_1, a_2, \ldots, a_{2l}\} = \text{sort } \{b_1^{(1)}, b_2^{(1)}, \ldots, b_1^{(l)}, b_2^{(l)}\},
$$
$$
a_{i_1} \equiv a_1 = \cdots = a_{l_1},
$$
$$
a_{i_2} \equiv a_{l_1+1} = \cdots = a_{l_1+l_2},
$$
$$
\vdots
$$
$$
a_{i_r} \equiv a_{\sum_{i=1}^{r-1} l_i + 1} = \cdots = a_{\sum_{i=1}^{r-1} l_i + l_r} = a_{2l}
$$
$$
\text{with } \sum_{s=1}^{r} l_s = 2l,
$$

(5.44)

then, the corresponding standard functional form is

$$
\dot{x} = a_0 \prod_{s=1}^{r} (g(x) - a_{i_s})^{l_s} \prod_{k=l+1}^{n} [(x + \frac{1}{2} B_{i_k})^2 + \frac{1}{4}(-\Delta_{i_k})]^{q_{i_k}}.
$$

(5.45)

The functional equilibrium separatrix flow is called an $(l_1 \text{th XX} : l_2 \text{th XX} : \ldots : l_r \text{th XX})$-flow.

(a$_1$) For $a_0 dg/dx|_{x*} > 0$ $(p = 1, 2, \ldots, r)$,

$$
l_p \text{th XX} = \begin{cases} (2r_p - 1)^{\text{th}} \text{ order source, for } \alpha_p = 2M_p - 1, \ l_p = 2r_p - 1; \\ (2r_p - 1)^{\text{th}} \text{ order sink, for } \alpha_p = 2M_p, \ l_p = 2r_p - 1, \end{cases}
$$

(5.46)

where

$$\alpha_p = \sum_{s=p}^{r} l_s. \tag{5.47}$$

(a$_2$) For $a_0 dg/dx|_{x^*} < 0$ $(p = 1, 2, \ldots, r)$,

$$l_p \text{th XX} = \begin{cases} (2r_p)^{\text{th}} \text{ order sink, for } \alpha_p = 2M_p - 1, \, l_p = 2r_p - 1; \\ (2r_p)^{\text{th}} \text{ order order source, for } \alpha_p = 2M_p, \, l_p = 2r_p - 1. \end{cases} \tag{5.48}$$

(a$_3$) For $a_0 > 0$ and $dg/dx|_{x^*} \neq 0$ $(p = 1, 2, \ldots, r)$,

$$l_p \text{th XX} = \begin{cases} (2r_p)^{\text{th}} \text{ order lower-saddle, for } \alpha_p = 2M_p - 1, \, l_p = 2r_p; \\ (2r_p)^{\text{th}} \text{ order upper-saddle, for } \alpha_p = 2M_p, \, l_p = 2r_p. \end{cases} \tag{5.49}$$

(a$_4$) For $a_0 < 0$ and $dg/dx|_{x^*} \neq 0$ $(p = 1, 2, \ldots, r)$,

$$l_p \text{th XX} = \begin{cases} (2r_p)^{\text{th}} \text{ order upper-saddle, for } \alpha_p = 2M_p - 1, \, l_p = 2r_p; \\ (2r_p)^{\text{th}} \text{ order lower-saddle, for } \alpha_p = 2M_p, \, l_p = 2r_p. \end{cases} \tag{5.50}$$

(b) The functional equilibrium of $x^* = a_{i_p}^{(s)} \in S_{i_p}$ with $g(x^*) = a_{i_p}$ for $(l_p > 1)$-repeated functional equilibriums switching $(p = 1, 2, \ldots, r)$ is called an l_pth XX bifurcation of $(l_{p_1}\text{th XX} : l_{p_2}\text{th XX} : \ldots : l_{p_\beta}\text{th XX})$-functional equilibrium switching at a point $\mathbf{p} = \mathbf{p}_1 \in \partial\Omega_{12}$, and the functional bifurcation condition is

$$a_{i_p} \equiv a_{\sum_{i=1}^{p-1} l_i + 1} = \cdots = a_{\sum_{i=1}^{p-1} l_i + l_p},$$

$$a_{\sum_{i=1}^{p-1} l_i + 1}^{\pm} \neq \cdots \neq a_{\sum_{i=1}^{p-1} l_i + l_p}^{\pm}; \tag{5.51}$$

$$l_p = \sum_{i=1}^{\beta} l_{p_i}, \quad p = 1, 2, \ldots, r.$$

(iii) If

$$\Delta_i = B_i^2 - 4C_i = 0, \, i \in \{i_{11}, i_{12}, \ldots, i_{1_s}\} \subseteq \{i_1, i_2, \ldots, i_l\} \subseteq \{1, 2, \ldots, n\},$$
$$\Delta_k = B_k^2 - 4C_k > 0, \, k \in \{i_{21}, i_{22}, \ldots, i_{2r}\} \subseteq \{i_1, i_2, \ldots, i_l\} \subseteq \{1, 2, \ldots, n\},$$
$$\Delta_j = B_j^2 - 4C_j < 0, \, j \in \{i_{l+1}, i_{l+2}, \ldots, i_n\} \subseteq \{1, 2, \ldots, n\} \text{ with } i \neq j \neq k,$$
$$\tag{5.52}$$

the 1-dimensional, $(2m)$th-degree polynomial functional dynamical system has $2l$-equilibriums as

$$g(x^*) = b_1^{(i)} = -\frac{1}{2}B_i, \ g(x^*) = b_2^{(i)} = -\frac{1}{2}B_i,$$

$$\text{for } i \in \{i_{11}, i_{12}, \ldots, i_{ls}\},$$

$$g(x^*) = b_1^{(k)} = -\frac{1}{2}(B_k + \sqrt{\Delta_k}),$$

$$g(x^*) = b_2^{(k)} = -\frac{1}{2}(B_k - \sqrt{\Delta_k}) \tag{5.53}$$

$$\text{for } k \in \{i_{21}, i_{22}, \ldots, i_{2r}\};$$

$$\{a_1, a_2, \ldots, a_{2l}\} = \text{sort}\,\{b_1^{(1)}, b_2^{(1)}, \ldots, b_1^{(l)}, b_2^{(l)}\}, \ a_s < a_{s+1};$$

$$S_\alpha = \{a_\alpha^{(s_\alpha)} | g(a_\alpha^{(s_\alpha)}) = a_\alpha, \ s_\alpha = 1, 2, \ldots, N_\alpha\} \cup \{\emptyset\},$$

$$\alpha = \{1, 2, \ldots, 2l\}.$$

If

$$\{a_1, a_2, \ldots, a_{2l}\} = \text{sort}\,\{b_1^{(1)}, b_2^{(1)}, \ldots, b_1^{(l)}, b_2^{(l)}\},$$

$$a_{i_1} \equiv a_1 = \cdots = a_{l_1},$$

$$a_{i_2} \equiv a_{l_1+1} = \cdots = a_{l_1+l_2},$$

$$\vdots \tag{5.54}$$

$$a_{i_r} \equiv a_{\sum_{i=1}^{r-1} l_i + 1} = \cdots = a_{\sum_{i=1}^{r-1} l_i + l_r} = a_{2l}$$

$$\text{with } \sum_{s=1}^{r} l_s = 2l,$$

then the corresponding standard functional form is

$$\dot{x} = a_0 \prod_{s=1}^{r} (g(x) - a_{i_s})^{l_s} \prod_{k=l+1}^{n} [(g(x) + \frac{1}{2}B_{i_k})^2 + \frac{1}{4}(-\Delta_{i_k})]^{q_{i_k}}. \tag{5.55}$$

The functional equilibrium separatrix flow is called an $(l_1\text{th XX}:l_2\text{th XX}:\ldots:l_r\text{th XX})$-flow.

(a) The functional equilibrium of $x^* = a_{i_p}^{(s)} \in S_{i_p}$ with $g(x^*) = a_{i_p}$ for $(l_p > 1)$-repeated functional equilibriums appearance (or vanishing) $(p = 1, 2, \ldots, r)$ is called an l_pth XX bifurcation of functional equilibrium at a point $\mathbf{p} = \mathbf{p}_1 \in \partial\Omega_{12}$, and the functional bifurcation condition is

$$a_{i_p} \equiv a_{\sum_{i=1}^{p-1} l_i+1} = \cdots = a_{\sum_{i=1}^{p-1} l_i+l_p} = -\frac{1}{2} B_{i_q},$$

$$\text{with } \Delta_{i_q} = B_{i_q}^2 - 4C_{i_q} = 0 \, (i_q \in \{i_1, i_2, \ldots i_l\}), \tag{5.56}$$

$$a^+_{\sum_{i=1}^{p-1} l_i+1} \neq \cdots \neq a^+_{\sum_{i=1}^{p-1} l_i+l_p} \text{ or } a^-_{\sum_{i=1}^{p-1} l_i+1} \neq \cdots \neq a^-_{\sum_{i=1}^{p-1} l_i+l_p}.$$

(b) The functional equilibrium of $x^* = a_{i_q}^{(s)} \in S_{i_q}$ with $g(x^*) = a_{i_q}$ for $(l_q > 1)$-repeated equilibriums switching $(q = 1, 2, \ldots, r_1)$ is called an l_qth XX bifurcation of $(l_{q_1}$th XX$: l_{q_2}$th XX$: \ldots : l_{q_\beta}$th XX$)$-functional equilibrium switching at a point $\mathbf{p} = \mathbf{p}_1 \in \partial\Omega_{12}$, and the bifurcation condition is

$$a_{i_q} \equiv a_{\sum_{i=1}^{q-1} l_i+1} = \cdots = a_{\sum_{i=1}^{q-1} l_i+l_q},$$

$$a^{\pm}_{\sum_{i=1}^{q-1} l_i+1} \neq \cdots \neq a^{\pm}_{\sum_{i=1}^{q-1} l_i+l_q}; \tag{5.57}$$

$$l_q = \sum_{i=1}^{\beta} l_{q_i}, \, q = 1, 2, \ldots, r_1.$$

(iv) If

$$\Delta_i = B_i^2 - 4C_i > 0 \text{ for } i = 1, 2, \ldots, n \tag{5.58}$$

the 1-dimensional, $(2m)$th-degree polynomial system has $(2n)$-equilibriums as

$$S_\alpha = \{a_\alpha^{(s_\alpha)} | g(a_\alpha^{(s_\alpha)}) = a_\alpha, \, s_\alpha = 1, 2, \ldots, N_\alpha\} \cup \{\emptyset\},$$

$$\alpha = \{1, 2, \ldots, 2n\};$$

$$g(x^*) = b_1^{(i)} = -\frac{1}{2}(B_i + \sqrt{\Delta_i}), \tag{5.59}$$

$$g(x^*) = b_2^{(i)} = -\frac{1}{2}(B_i - \sqrt{\Delta_i}),$$

$$\text{for } i = 1, 2, \ldots, n.$$

(iv$_1$) If

$$b_r^{(i)} \neq b_s^{(j)} \text{ for } r, s \in \{1, 2\}; \, i, j = 1, 2, \ldots, n$$

$$\{a_1, a_2, \ldots, a_{2n}\} = \text{sort} \, \{b_1^{(1)}, b_2^{(1)}, \ldots, b_1^{(n)}, b_2^{(n)}\}, \, a_s < a_{s+1}. \tag{5.60}$$

The corresponding standard functional form is

$$\dot{x} = a_0 \prod_{s=1}^{2n} (g(x) - a_s)^{l_s} \text{ with } l_s \in \{q_{i_1}, q_{i_2}, \ldots, q_{i_n}\}. \tag{5.61}$$

The functional equilibrium separatrix flow is called an $(l_1$th XX$: l_2$th XX$: \ldots : l_{2n}$th XX$)$-flow.

(a_1) For $a_0 dg/dx|_{x^*} > 0$ $(p = 1, 2, \ldots, 2n)$,

$$
l_p \text{th XX} = \begin{cases} (2r_p - 1)^{\text{th}} \text{ order source, for } \alpha_p = 2M_p - 1, \, l_p = 2r_p - 1; \\ (2r_p - 1)^{\text{th}} \text{ order sink, for } \alpha_p = 2M_p, \, l_p = 2r_p - 1, \end{cases}
$$

(5.62)

where

$$
\alpha_p = \sum_{s=p}^{2n} l_s. \tag{5.63}
$$

(a_2) For $a_0 dg/dx|_{x^*} < 0$ $(p = 1, 2, \ldots, 2n)$,

$$
l_p \text{th XX} = \begin{cases} (2r_p - 1)^{\text{th}} \text{ order sink, for } \alpha_p = 2M_p - 1, \, l_p = 2r_p - 1; \\ (2r_p - 1)^{\text{th}} \text{ order source, for } \alpha_p = 2M_p, \, l_p = 2r_p - 1. \end{cases}
$$

(5.64)

(a_3) For $a_0 > 0$ and $dg/dx|_{x^*} \neq 0$ $(p = 1, 2, \ldots, 2n)$,

$$
l_p \text{th XX} = \begin{cases} (2r_p)^{\text{th}} \text{ order lower-saddle, for } \alpha_p = 2M_p - 1, \, l_p = 2r_p; \\ (2r_p)^{\text{th}} \text{ order upper-ssaddle, for } \alpha_p = 2M_p, \, l_p = 2r_p 1. \end{cases}
$$

(5.65)

(a_4) For $a_0 < 0$ and $dg/dx|_{x^*} \neq 0$ $(p = 1, 2, \ldots, 2n)$,

$$
l_p \text{th XX} = \begin{cases} (2r_p)^{\text{th}} \text{ order upper-saddle, for } \alpha_p = 2M_p - 1, \, l_p = 2r_p; \\ (2r_p)^{\text{th}} \text{ order lower-saddle, for } \alpha_p = 2M_p, \, l_p = 2r_p. \end{cases}
$$

(5.66)

(iv_2) If

$$
\{a_1, a_2, \ldots, a_{2n}\} = \text{sort } \{b_1^{(1)}, b_2^{(1)}, \ldots, b_1^{(n)}, b_2^{(n)}\},
$$

$$
a_{i_1} \equiv a_1 = \cdots = a_{l_1},
$$

$$
a_{i_2} \equiv a_{l_1+1} = \cdots = a_{l_1+l_2},
$$

$$
\vdots
$$

(5.67)

$$
a_{i_r} \equiv a_{\sum_{i=1}^{r-1} l_i + 1} = \cdots = a_{\sum_{i=1}^{r-1} l_i + l_r} = a_{2n},
$$

$$
\text{with } \sum_{s=1}^{r} l_s = 2n,
$$

then the corresponding standard functional form is

$$
\dot{x} = a_0 \prod_{s=1}^{r} (g(x) - a_{i_s})^{l_s}. \tag{5.68}
$$

The functional equilibrium separatrix flow is called an $(l_1 \text{th XX} : l_2 \text{th XX} : \ldots : l_r \text{th XX})$-flow. The functional equilibrium of $x^* = a_{i_p}^{(s)} \in S_{i_p}$ with $g(x^*) = a_{i_p}$ for l_p-repeated functional equilibriums switching $(p = 1, 2, \ldots, r)$ is called an $l_p \text{th XX}$ bifurcation of $(l_{p_1} \text{th XX} : l_{p_2} \text{th XX} : \ldots : l_{p_\beta} \text{th XX})$-functional equilibrium switching at a point $\mathbf{p} = \mathbf{p}_1 \in \partial\Omega_{12}$, and the functional bifurcation condition is

$$a_{i_p} \equiv a_{\sum_{i=1}^{p-1} l_i + 1} = \cdots = a_{\sum_{i=1}^{p-1} l_i + l_p},$$

$$a^\pm_{\sum_{i=1}^{p-1} l_i + 1} \neq \cdots \neq a^\pm_{\sum_{i=1}^{p-1} l_i + l_p};$$

(5.69)

$$l_p = \sum_{i=1}^{\beta} l_{p_i}, \quad p = 1, 2, \ldots, r.$$

Definition 5.3   Consider a 1-dimensional, $(2m)$th-degree polynomial nonlinear functional dynamical system

$$\dot{x} = A_0(\mathbf{p})(g(x))^{2m} + A_1(\mathbf{p})(g(x))^{2m-1} +$$
$$\cdots + A_{2m-2}(\mathbf{p})(g(x))^2 + A_{2m-1}g(x) + A_{2m}(\mathbf{p})$$
$$= a_0(\mathbf{p}) \prod_{s=1}^{r} (g(x) - c_s(\mathbf{p}))^{l_s} \prod_{i=r+1}^{n} [(g(x))^2 + B_i(\mathbf{p})g(x) + C_i(\mathbf{p})]^{q_i},$$

(5.70)

where $A_0(\mathbf{p}) \neq 0$, and

$$\sum_{s=1}^{r} l_s = 2l, \quad \sum_{i=r+1}^{n} q_i = (m - l), \quad \mathbf{p} = (p_1, p_2, \ldots, p_{m_1})^{\mathrm{T}}.$$

(5.71)

(i) If

$$\Delta_i = B_i^2 - 4C_i < 0 \text{ for } i = r + 1, r + 2, \ldots, n,$$
$$\{a_1, a_2, \ldots, a_r\} = \text{sort}\{c_1, c_2, \ldots, c_r\}, \text{ with } a_i < a_{i+1},$$
$$S_\alpha = \{a_\alpha^{(s_\alpha)} | g(a_\alpha^{(s_\alpha)}) = a_\alpha, \ s_\alpha = 1, 2, \ldots, N_\alpha\} \cup \{\emptyset\},$$
$$\alpha \in \{1, 2, \ldots, r\};$$

(5.72)

the 1-dimensional nonlinear functional dynamical system with a $(2m)$th-degree polynomial have equilibriums of $x^* = a_\alpha^{(s_\alpha)} \in S_\alpha$ with $g(x^*) = a_\alpha$ $(\alpha = 1, 2, \ldots, r)$, and the corresponding standard functional form is

$$\dot{x} = a_0(\mathbf{p}) \prod_{s=1}^{r} (g(x) - a_{i_s})^{l_s} \prod_{i=r+1}^{n} [(g(x) + \frac{1}{2}B_i)^2 + \frac{1}{4}(-\Delta_i)]^{l_i}.$$

(5.73)

The functional equilibrium separatrix flow is called an $(l_1\text{th XX}:l_2\text{th XX}:\ldots:l_r\text{th XX})$-flow.

(a₁) For $a_0 dg/dx|_{x^*} > 0$ $(s = 1, 2, \ldots, r)$,

$$l_p\text{th XX} = \begin{cases} (2r_p - 1)^{\text{th}} \text{ order source, for } \alpha_p = 2M_p - 1, \, l_p = 2r_p - 1; \\ (2r_p - 1)^{\text{th}} \text{ order sink, for } \alpha_p = 2M_p, \, l_p = 2r_p - 1, \end{cases}$$

(5.74)

where

$$\alpha_p = \sum_{s=p}^{r} l_s.$$

(5.75)

(a₂) For $a_0 dg/dx|_{x^*} < 0$ $(s = 1, 2, \ldots, r)$,

$$l_p\text{th XX} = \begin{cases} (2r_p - 1)^{\text{th}} \text{ order sink, for } \alpha_p = 2M_p - 1, \, l_p = 2r_p - 1; \\ (2r_p - 1)^{\text{th}} \text{ order source, for } \alpha_p = 2M_p, \, l_p = 2r_p - 1. \end{cases}$$

(5.76)

(a₃) For $a_0 > 0$ and $dg/dx|_{x^*} \neq 0$ $(p = 1, 2, \ldots, r)$,

$$l_p\text{th XX} = \begin{cases} (2r_p)^{\text{th}} \text{ order lower-saddle, for } \alpha_p = 2M_p - 1, \, l_p = 2r_p; \\ (2r_p)^{\text{th}} \text{ order upper-saddle, for } \alpha_p = 2M_p, \, l_p = 2r_p. \end{cases}$$

(5.77)

(a₄) For $a_0 < 0$ and $dg/dx|_{x^*} \neq 0$ $(p = 1, 2, \ldots, r)$,

$$l_p\text{th XX} = \begin{cases} (2r_p)^{\text{th}} \text{ order upper-saddle, for } \alpha_p = 2M_p - 1, \, l_p = 2r_p; \\ (2r_p)^{\text{th}} \text{ order lower-saddle, for } \alpha_p = 2M_p, \, l_p = 2r_p. \end{cases}$$

(5.78)

(ii) If

$$\begin{aligned} \Delta_i &= B_i^2 - 4C_i > 0, \, i = j_1, j_2, \ldots, j_s \in \{l + 1, l + 2, \ldots, n\}, \\ \Delta_j &= B_j^2 - 4C_j < 0, \, j = j_{s+1}, j_{s+2}, \ldots, j_n \in \{l + 1, l + 2, \ldots, n\} \\ &\text{with } s \in \{1, \ldots, n - l\}, \end{aligned}$$

(5.79)

the 1-dimensional, $(2m)$th-degree polynomial functional system has $2n_2$-functional equilibriums as

$$g(x^*) = b_1^{(i)} = -\frac{1}{2}(B_i + \sqrt{\Delta_i}),$$

$$g(x^*) = b_2^{(i)} = -\frac{1}{2}(B_i - \sqrt{\Delta_i})$$

$$i \in \{j_1, j_2, \dots, j_{n_1}\} \subseteq \{l+1, l+2, \dots, n\};$$

$$\{a_1, a_2, \dots, a_{2n_2}\} = \text{sort}\{c_1, c_2, \dots, c_{2l}, \underbrace{b_1^{(r+l)}, b_2^{(r+1)}}_{q_{r+1} \text{ sets}}, \dots, \underbrace{b_1^{(n_1)}, b_2^{(n_1)}}_{q_{n_1} \text{ sets}}\};$$

$$S_\alpha = \{a_\alpha^{(s_\alpha)} | g(a_\alpha^{(s_\alpha)}) = a_\alpha, \ s_\alpha = 1, 2, \dots, N_\alpha\} \cup \{\emptyset\},$$

$$\alpha \in \{1, 2, \dots, r\}.$$

(5.80)

If

$$\{a_1, a_2, \dots, a_{2n_2}\} = \text{sort}\{c_1, c_2, \dots, c_{2l}, \underbrace{b_1^{(r+1)}, b_2^{(r+1)}}_{q_{r+1} \text{ sets}}, \dots, \underbrace{b_1^{(n_1)}, b_2^{(n_1)}}_{q_{n_1} \text{ sets}}\};$$

$$a_{i_1} \equiv a_1 = \dots = a_{l_1},$$

$$a_{i_2} \equiv a_{l_1+1} = \dots = a_{l_1+l_2},$$

$$\vdots$$

$$a_{i_{n_1}} \equiv a_{\sum_{i=1}^{n_1-1} l_i + 1} = \dots = a_{\sum_{i=1}^{n_1-1} l_i + l_{n_1}} = a_{2n_2}$$

$$\text{with } \sum_{s=1}^{n_1} l_s = 2n_2,$$

(5.81)

then, the corresponding standard functional form is

$$\dot{x} = a_0 \prod_{s=1}^{n_1} (g(x) - a_{i_s})^{l_s} \prod_{i=n_2+1}^{n} [(g(x) + \frac{1}{2}B_i)^2 + \frac{1}{4}(-\Delta_i)]^{q_i}.$$

(5.82)

The functional equilibrium separatrix flow is called an $(l_1\text{th}\,XX : l_2\text{th}\,XX : \dots : l_{n_1}\text{th}\,XX)$-flow.

(a_1) For $a_0 dg/dx|_{x^*} > 0$ $(p = 1, 2, \dots, r, r+1, \dots, n_1)$,

$$l_p\text{th}\,XX = \begin{cases} (2r_p - 1)^{\text{th}} \text{ order source, for } \alpha_p = 2M_p - 1, \ l_p = 2r_p - 1; \\ (2r_p - 1)^{\text{th}} \text{ order sink, for } \alpha_p = 2M_p, \ l_p = 2r_p - 1, \end{cases}$$

(5.83)

where

$$\alpha_p = \sum_{s=p}^{n_1} l_s. \tag{5.84}$$

(a_2) For $a_0 dg/dx|_{x^*} < 0$ $(p = 1, 2, \ldots, r, r+1, \ldots, n_1)$,

$$l_p \text{th XX} = \begin{cases} (2r_p - 1)^{\text{th}} \text{ order sink, for } \alpha_p = 2M_p - 1, \ l_p = 2r_p - 1; \\ (2r_p - 1)^{\text{th}} \text{ order source, for } \alpha_p = 2M_p, \ l_p = 2r_p - 1. \end{cases} \tag{5.85}$$

(a_3) For $a_0 > 0$ and $dg/dx|_{x^*} \neq 0$ $(p = 1, 2, \ldots, r, r+1, \ldots, n_1)$,

$$l_p \text{th XX} = \begin{cases} (2r_p)^{\text{th}} \text{ order lower-saddle, for } \alpha_p = 2M_p - 1, \ l_p = 2r_p; \\ (2r_p)^{\text{th}} \text{ order upper-saddle, for } \alpha_p = 2M_p, \ l_p = 2r_p. \end{cases} \tag{5.86}$$

(a_4) For $a_0 < 0$ and $dg/dx|_{x^*} \neq 0$ $(p = 1, 2, \ldots, r, r+1, \ldots, n_1)$,

$$l_p \text{th XX} = \begin{cases} (2r_p)^{\text{th}} \text{ order upper-saddle, for } \alpha_p = 2M_p - 1, \ l_p 2r_p; \\ (2r_p)^{\text{th}} \text{ order lower-saddle, for } \alpha_p = 2M_p, \ l_p = 2r_p. \end{cases} \tag{5.87}$$

(b) The functional equilibrium of $x^* = a_{i_p}^{(s)} \in S_{i_p}$ with $g(x^*) = a_{i_p}$ for $(l_p > 1)$-repeated functional equilibriums switching is called an l_pth XX switching bifurcation of $(l_{p_1} \text{th XX} : l_{p_2} \text{th XX} : \ldots : l_{p_\beta} \text{th XX})$ functional equilibrium at a point $\mathbf{p} = \mathbf{p}_1 \in \partial\Omega_{12}$, and the functional bifurcation condition is

$$\begin{aligned} a_{i_p} &\equiv a_{\sum_{i=1}^{p-1} l_i + 1} = \cdots = a_{\sum_{i=1}^{p-1} l_i + l_p}, \\ a^{\pm}_{\sum_{i=1}^{p-1} l_i + 1} &\neq \cdots \neq a^{\pm}_{\sum_{i=1}^{p-1} l_i + l_p}, \\ p &= 1, 2, \ldots, n_1. \end{aligned} \tag{5.88}$$

(iii) If

$$\begin{aligned} \Delta_i &= B_i^2 - 4C_i = 0, \\ &\quad \text{for } i \in \{i_{11}, i_{12}, \ldots, i_{1_s}\} \subseteq \{i_{l+1}, i_{l+2}, \ldots, i_{n_2}\} \subseteq \{l+1, l+2, \ldots, n\}, \\ \Delta_k &= B_k^2 - 4C_k > 0, \\ &\quad \text{for } k \in \{i_{21}, i_{22}, \ldots, i_{2r}\} \subseteq \{i_{l+1}, i_{l+2}, \ldots, i_{n_2}\} \subseteq \{l+1, l+2, \ldots, n\}, \\ \Delta_j &= B_j^2 - 4C_j < 0, \\ &\quad \text{for } j \in \{i_{n_2+1}, i_{n_2+2}, \ldots, i_n\} \subseteq \{l+1, l+2, \ldots, n\}, \end{aligned} \tag{5.89}$$

the 1-dimensional, $(2m)$th-degree polynomial system has $2n_2$-equilibriums as

$$g(x^*) = b_1^{(i)} = -\frac{1}{2}B_i, \ g(x^*) = b_2^{(i)} = -\frac{1}{2}B_i,$$

$$\text{for } i \in \{i_{11}, i_{12}, \dots, i_{1_s}\},$$

$$g(x^*) = b_1^{(k)} = -\frac{1}{2}(B_k + \sqrt{\Delta_k}),$$

$$g(x^*) = b_2^{(k)} = -\frac{1}{2}(B_k - \sqrt{\Delta_k}) \tag{5.90}$$

$$\text{for } k \in \{i_{21}, i_{22}, \dots, i_{2r}\};$$

$$\{a_1, a_2, \dots, a_{2n_2}\} = \text{sort} \{c_1, c_2, \dots, c_{2l}, \underbrace{b_1^{(r+1)}, b_2^{(r+1)}}_{q_{r+1} \text{ sets}}, \dots, \underbrace{b_1^{(n_1)}, b_2^{(n_1)}}_{q_{n_1} \text{ sets}}\},$$

$$S_\alpha = \{a_\alpha^{(s_\alpha)} | g(a_\alpha^{(s_\alpha)}) = a_\alpha, \ s_\alpha = 1, 2, \dots, N_\alpha\} \cup \{\emptyset\},$$

$$\alpha \in \{1, 2, \dots, 2n_2\}.$$

If

$$\{a_1, a_2, \dots, a_{2n_2}\} = \text{sort} \{c_1, c_2, \dots, c_{2l}, \underbrace{b_1^{(r+1)}, b_2^{(r+1)}}_{q_{r+1} \text{ sets}}, \dots, \underbrace{b_1^{(n_1)}, b_2^{(n_1)}}_{q_{n_1} \text{ sets}}\},$$

$$a_{i_1} \equiv a_1 = \dots = a_{l_1},$$

$$a_{i_2} \equiv a_{l_1+1} = \dots = a_{l_1+l_2},$$

$$\vdots \tag{5.91}$$

$$a_{i n_1} \equiv a_{\sum_{i=1}^{n_1-1} l_i + 1} = \dots = a_{\sum_{i=1}^{n_1-1} l_i + l_{n_1}} = a_{2n_2}$$

$$\text{with } \sum_{s=1}^{n_1} l_s = 2n_2,$$

then, the corresponding standard functional form is

$$\dot{x} = a_0 \prod_{s=1}^{r_1} (g(x) - a_{i_s})^{l_s} \prod_{i=n_2+1}^{n} [(g(x) + \frac{1}{2}B_i)^2 + \frac{1}{4}(-\Delta_i)]^{q_i}. \tag{5.92}$$

The equilibrium separatrix flow is called an $(l_1 \text{th XX} : l_2 \text{th XX} : \dots : l_{r_1} \text{th XX})$-flow.

(a) The functional equilibrium of $x^* = a_{i_p}^{(s)} \in S_{i_p}$ with $g(x^*) = a_{i_p}$ for $(l_p > 1)$-repeated functional equilibriums appearance or vanishing is called an l_pth XX bi-furcation of functional equilibrium at a point $\mathbf{p} = \mathbf{p}_1 \in \partial\Omega_{12}$, and the functional

bifurcation condition is

$$a_{i_p} \equiv a_{\sum_{i=1}^{p-1} l_i + 1} = \cdots = a_{\sum_{i=1}^{p-1} l_i + l_p} = -\frac{1}{2} B_{i_q},$$
$$\text{with } \Delta_{i_q} = B_{i_q}^2 - 4C_{i_q} = 0 \, (i_q \in \{i_1, i_2, \ldots i_l\}), \tag{5.93}$$
$$a_{\sum_{i=1}^{p-1} q_i + 1}^+ \neq \cdots \neq a_{\sum_{i=1}^{p-1} q_i + q_p}^+ \quad \text{or} \quad a_{\sum_{i=1}^{p-1} q_i + 1}^- \neq \cdots \neq a_{\sum_{i=1}^{p-1} q_i + q_p}^-.$$

(b) The functional equilibrium of $x^* = a_{i_p}^{(s)} \in S_{i_p}$ with $g(x^*) = a_{i_p}$ for $(l_p > 1)$-repeated functional equilibriums switching is called an l_pth XX bifurcation of $(l_{p_1}$ th XX : l_{p_2} th XX : ... : l_{p_β} th XX) functional equilibrium switching at a point $\mathbf{p} = \mathbf{p}_1 \in \partial\Omega_{12}$, and the bifurcation condition is

$$a_{i_p} \equiv a_{\sum_{i=1}^{p-1} l_i + 1} = \cdots = a_{\sum_{i=1}^{p-1} l_i + l_p},$$
$$a_{\sum_{i=1}^{p-1} l_i + 1}^{\pm} \neq \cdots \neq a_{\sum_{i=1}^{p-1} l_i + l_p}^{\pm}, \tag{5.94}$$
$$l_p = \sum_{i=1}^{\beta} l_{p_i}, \quad p = 1, 2, \ldots, r_1.$$

(c) The equilibrium of $x^* = a_{i_p}^{(s)} \in S_{i_p}$ with $g(x^*) = a_{i_p}$ for $(l_{p_1} \geq 1)$-repeated functional equilibriums appearance (or vanishing) and $(l_{p_2} \geq 2)$ repeated functional equilibriums switching of $(l_{p_{21}}$ th XX : $l_{p_{22}}$ th XX : ... : $l_{p_{2\beta}}$ th XX) is called an l_pth XX bifurcation of functional equilibrium at a point $\mathbf{p} = \mathbf{p}_1 \in \partial\Omega_{12}$, and the functional bifurcation condition is

$$a_{i_p} \equiv a_{\sum_{i=1}^{p-1} l_i + 1} = \cdots = a_{\sum_{i=1}^{p-1} l_i + q_p}$$
$$\text{with } \Delta_{i_q} = B_{i_q}^2 - 4C_{i_q} = 0 \, (i_q \in \{i_1, i_2, \ldots, i_l\})$$
$$a_{\sum_{i=1}^{p-1} l_i + j_1}^+ \neq \cdots \neq a_{\sum_{i=1}^{p-1} l_i + j_{p_1}}^+ \quad \text{or} \quad a_{\sum_{i=1}^{p_1-1} l_i + j_1}^- \neq \cdots \neq a_{\sum_{i=1}^{p_1-1} l_i + j_{p_1}}^-,$$
$$\text{for } \{j_1, j_2, \ldots, j_{p_1}\} \subseteq \{1, 2, \ldots, l_p\},$$
$$a_{\sum_{i=1}^{p-1} l_i + k_1}^{\pm} \neq \cdots \neq a_{\sum_{i=1}^{p-1} l_i + k_{p_2}}^{\pm}$$
$$\text{for } \{k_1, k_2, \ldots, k_{p_2}\} \subseteq \{1, 2, \ldots, l_p\},$$
$$\text{with } l_{p_1} + l_{p_2} = l_p, \quad p = 1, 2, \ldots, r_1. \tag{5.95}$$

(iv) If

$$\Delta_i = B_i^2 - 4C_i > 0 \text{ for } i = l + 1, l + 2, \ldots, n, \tag{5.96}$$

the 1-dimensional, $(2m)$th-degree polynomial functional system has $(2m)$ functional equilibriums as

$$g(x^*) = b_1^{(i)} = -\frac{1}{2}(B_i + \sqrt{\Delta_i}),$$

$$g(x^*) = b_2^{(i)} = -\frac{1}{2}(B_i - \sqrt{\Delta_i})$$

for $i = l + 1, l + 2, \ldots, n;$

$$\{a_1, a_2, \ldots, a_{2m}\} = \text{sort}\{c_1, c_2, \ldots, c_{2l}, \underbrace{b_1^{(r+1)}, b_2^{(r+1)}}_{q_{r+1} \text{ sets}}, \ldots, \underbrace{b_1^{(n)}, b_2^{(n)}}_{q_n \text{ sets}}\},$$ (5.97)

$$S_\alpha = \{a_\alpha^{(s_\alpha)} | g(a_\alpha^{(s_\alpha)}) = a_\alpha, \ s_\alpha = 1, 2, \ldots, N_\alpha\} \cup \{\emptyset\},$$

$$\alpha \in \{1, 2, \ldots, 2m\}.$$

If

$$\{a_1, a_2, \ldots, a_{2m}\} = \text{sort}\{c_1, c_2, \ldots, c_{2l}, \underbrace{b_1^{(r+1)}, b_2^{(r+1)}}_{q_{r+1} \text{ sets}}, \ldots, \underbrace{b_1^{(n)}, b_2^{(n)}}_{q_n \text{ sets}}\},$$

$$a_{i_1} \equiv a_1 = \cdots = a_{l_1},$$

$$a_{i_2} \equiv a_{l_1+1} = \cdots = a_{l_1+l_2},$$

$$\vdots$$ (5.98)

$$a_{i_n} \equiv a_{\sum_{i=1}^{n-1} l_i + 1} = \cdots = a_{\sum_{i=1}^{n-1} l_i + l_r} = a_{2m}$$

$$\text{with } \sum_{s=1}^{n} l_s = 2m,$$

then, the corresponding standard functional form is

$$\dot{x} = a_0 \prod_{s=1}^{n} (g(x) - a_{i_s})^{l_s}.$$ (5.99)

The functional equilibrium separatrix flow is called an $(l_1\text{th XX} : l_2\text{th XX} : \ldots : l_n\text{th XX})$-flow. The functional equilibrium of $x^* = a_{i_p}^{(s)} \in S_{i_p}$ with $g(x^*) = a_{i_p}$ for l_p-repeated functional equilibriums switching is called an l_pth XX switching bifurcation of $(l_{p_1}\text{th XX} : l_{p_2}\text{th XX} : \ldots : l_{p_\beta}\text{th XX})$ functional equilibrium at a point $\mathbf{p} = \mathbf{p}_1 \in \partial\Omega_{12}$, and the functional bifurcation condition is

$$a_{i_p} \equiv a_{\sum_{i=1}^{p-1} l_i + 1} = \cdots = a_{\sum_{i=1}^{p-1} l_i + l_p},$$

$$a^{\pm}_{\sum_{i=1}^{p-1} l_i + 1} \neq \cdots \neq a^{\pm}_{\sum_{i=1}^{p-1} l_i + l_p};$$

$$l_p = \sum_{i=1}^{\beta} l_{p_i}, \quad p = 1, 2, \ldots, n. \tag{5.100}$$

CHAPTER 6

(2*m*+1)th-Degree Polynomial Functional Systems

In this chapter, the global stability and bifurcation of functional equilibriums in the $(2m + 1)$th-degree polynomial nonlinear functional dynamical systems.

Definition 6.1   Consider a $(2m + 1)$th-degree polynomial nonlinear functional dynamical system

$$
\begin{aligned}
\dot{x} &= A_0(\mathbf{p})(g(x))^{2m+1} + A_1(\mathbf{p})(g(x))^{2m} + \\
&\quad \cdots + A_{2m-1}(\mathbf{p})(g(x))^2 + A_{2m}g(x) + A_{2m+1}(\mathbf{p}) \\
&= a_0(\mathbf{p})(g(x) - a(\mathbf{p}))[(g(x))^2 + B_1(\mathbf{p})g(x) + C_1(\mathbf{p})] \\
&\quad \cdots [(g(x))^2 + B_m(\mathbf{p})g(x) + C_m(\mathbf{p})],
\end{aligned}
\tag{6.1}
$$

where $A_0(\mathbf{p}) \neq 0$, and

$$
\mathbf{p} = (p_1, p_2, \dots, p_m)^{\mathrm{T}}.
\tag{6.2}
$$

(i) If

$$
\begin{aligned}
&\Delta_i = B_i^2 - 4C_i < 0 \text{ for } i = 1, 2, \dots, m, \\
&S_1 = \{a_1^{(s)} | g(a_1^{(s_1)}) = a_1, \ s_1 = 1, 2, \dots, N_1\} \cup \{\emptyset\}, \\
&a_1 = a.
\end{aligned}
\tag{6.3}
$$

the $(2m + 1)$th-degree polynomial system has one equilibrium of $x^* = a_1^{(s)} \in S_1$ for $g(x^*) = a$, and the corresponding standard form is

$$
\begin{aligned}
\dot{x} &= a_0(g(x) - a)[(g(x) + \frac{1}{2}B_1)^2 + \frac{1}{4}(-\Delta_1)] \\
&\quad \cdots [(g(x) + \frac{1}{2}B_m)^2 + \frac{1}{4}(-\Delta_m)].
\end{aligned}
\tag{6.4}
$$

The flow of such a functional system with one functional equilibrium is called a single-functional equilibrium flow.

(a) If $a_0 dg/dx|_{x^*} > 0$, the functional equilibrium flow with $x^* = a_1^{(s)} \in S_1$ for $g(x^*) = a$ is called a functional source flow.

(b) If $a_0 dg/dx|_{x^*} < 0$, the functional equilibrium flow with $x^* = a_1^{(s)} \in S_1$ for $g(x^*) = a$ is called a functional sink flow.

(ii) If

$$\Delta_i = B_i^2 - 4C_i > 0, \ i = i_1, i_2, \dots, i_l \in \{1, 2, \dots, m\},$$
$$\Delta_j = B_j^2 - 4C_j < 0, \ j = i_{l+1}, i_{l+2}, \dots, i_m \in \{1, 2, \dots, m\} \tag{6.5}$$
$$\text{with } l \in \{0, 1, \dots, m\},$$

the $(2m + 1)$th-degree polynomial nonlinear functional dynamical system has $(2l + 1)$-functional equilibriums as

$$g(x^*) = b_1^{(i)} = -\frac{1}{2}(B_i + \sqrt{\Delta_i}),$$
$$g(x^*) = b_2^{(i)} = -\frac{1}{2}(B_i - \sqrt{\Delta_i}),$$
$$i \in \{i_1, i_2, \dots, i_l\} \subseteq \{1, 2, \dots, m\}; \tag{6.6}$$
$$\{a_1, a_2, \dots, a_{2l+1}\} = \text{sort}\{a, b_1^{(1)}, b_2^{(1)}, \dots, b_1^{(l)}, b_2^{(l)}\}, \ a_\alpha < a_{\alpha+1},$$
$$S_\alpha = \{a_\alpha^{(s_\alpha)} | g(a_\alpha^{(s_\alpha)}) = a_\alpha, \ s_\alpha = 1, 2, \dots, N_\alpha\} \cup \{\emptyset\},$$
$$\alpha = \{1, 2, \dots, 2l + 1\}.$$

(ii$_1$) If

$$b_r^{(i)} \neq b_s^{(j)} \text{ for } r, s \in \{1, 2\}; \ i, j = 1, 2, \dots, l;$$
$$\{a_1, a_2, \dots, a_{2l}\} = \text{sort}\{a, b_1^{(1)}, b_2^{(1)}, \dots, b_1^{(l)}, b_2^{(l)}\}, \ a_s < a_{s+1}, \tag{6.7}$$

then, the corresponding standard form is

$$\dot{x} = a_0 \prod_{i_1=1}^{2l+1} (g(x) - a_{i_1}) \prod_{k=l+1}^{m} [(g(x) + \frac{1}{2}B_{i_k})^2 + \frac{1}{4}(-\Delta_{i_k})]. \tag{6.8}$$

(a) If $a_0 dg/dx|_{x^*} > 0$, the simple functional equilibrium separatrix flow is called a functional (SO:SI:...:SO:SI:...:SI:SO) flow.

(b) If $a_0 dg/dx|_{x^*} < 0$, the simple functional equilibrium separatrix flow is called a functional (SI:SO:...:SI:SO:...:SO:SI) flow.

(ii$_2$) If

$$\{a_1, a_2, \ldots, a_{2l+1}\} = \text{sort}\{a, b_1^{(1)}, b_2^{(1)}, \ldots, b_1^{(l)}, b_2^{(l)}\},$$

$$a_{i_1} \equiv a_1 = \cdots = a_{l_1},$$

$$a_{i_2} \equiv a_{l_1+1} = \cdots = a_{l_1+l_2},$$

$$\vdots$$

$$a_{i_r} \equiv a_{\sum_{i=1}^{r-1} l_i + 1} = \cdots = a_{\sum_{i=1}^{r-1} l_i + l_r} = a_{2l+1}$$

$$\text{with } \sum_{s=1}^{r} l_s = 2l + 1,$$ (6.9)

then, the corresponding standard functional form is

$$\dot{x} = a_0 \prod_{s=1}^{r} (g(x) - a_{i_s})^{l_s} \prod_{k=l+1}^{m} [(g(x) + \frac{1}{2} B_{i_k})^2 + \frac{1}{4}(-\Delta_{i_k})]. \quad (6.10)$$

The functional equilibrium separatrix flow is called an $(l_1 \text{th XX} : l_2 \text{th XX} : \ldots : l_r \text{th XX})$-functional flow.

(a$_1$) For $a_0 dg/dx|_{x^*} > 0$ $(p = 1, 2, \ldots, r)$,

$$l_p \text{th XX} = \begin{cases} (2r_p - 1)^{\text{th}} \text{ order source, for } \alpha_p = 2M_p - 1, \ l_p = 2r_p - 1; \\ (2r_p - 1)^{\text{th}} \text{ order sink, for } \alpha_p = 2M_p, \ l_p = 2r_p - 1, \end{cases}$$

(6.11)

where

$$\alpha_p = \sum_{s=p}^{r} l_s. \quad (6.12)$$

(a$_2$) For $a_0 dg/dx|_{x^*} < 0$ $(p = 1, 2, \ldots, r)$,

$$l_p \text{th XX} = \begin{cases} (2r_p - 1)^{\text{th}} \text{ order sink, for } \alpha_p = 2M_p - 1, \ l_p = 2r_p - 1; \\ (2r_p - 1)^{\text{th}} \text{ order source, for } \alpha_p = 2M_p, \ l_p = 2r_p - 1. \end{cases}$$

(6.13)

(a$_3$) For $a_0 > 0$ and $dg/dx|_{x^*} \neq 0$ $(p = 1, 2, \ldots, r)$,

$$l_p \text{th XX} = \begin{cases} (2r_p)^{\text{th}} \text{ order lower-saddle, for } \alpha_p = 2M_p - 1, \ l_p = 2r_p; \\ (2r_p)^{\text{th}} \text{ order upper-saddle, for } \alpha_p = 2M_p, \ l_p = 2r_p. \end{cases}$$

(6.14)

(a$_4$) For $a_0 < 0$ and $dg/dx|_{x^*} \neq 0$ $(p = 1, 2, \ldots, r)$,

$$l_p \text{th XX} = \begin{cases} (2r_p)^{\text{th}} \text{ order upper-saddle, for } \alpha_p = 2M_p - 1, \ l_p = 2r_p; \\ (2r_p)^{\text{th}} \text{ order lower-saddle, for } \alpha_p = 2M_p, \ l_p = 2r_p. \end{cases}$$

(6.15)

(b) The functional equilibrium of $x^* = a_{i_p}^{(s)} \in S_{i_p}$ with $g(x^*) = a_{i_p}$ for $(l_p > 1)$-repeated functional equilibriums switching is called an l_pth XX bifurcation of $(l_{p_1}$th XX : l_{p_2}th XX : ... : l_{p_β}th XX$)$ functional equilibrium switching at a point $\mathbf{p} = \mathbf{p}_1 \in \partial\Omega_{12}$, and the functional bifurcation condition is

$$a_{i_p} \equiv a_{\sum_{i=1}^{p-1} l_i + 1} = \cdots = a_{\sum_{i=1}^{p-1} l_i + l_p},$$

$$a_{\sum_{i=1}^{p-1} l_i + 1}^{\pm} \neq \cdots \neq a_{\sum_{i=1}^{p-1} l_i + l_p}^{\pm};$$

$$l_p = \sum_{i=1}^{\beta} l_{p_i}, \quad p = 1, 2, \ldots, r. \tag{6.16}$$

(iii) If

$$\Delta_i = B_i^2 - 4C_i = 0, \ i \in \{i_{11}, i_{12}, \ldots, i_{l_s}\} \subseteq \{i_1, i_2, \ldots, i_l\} \subseteq \{1, 2, \ldots, m\},$$

$$\Delta_k = B_k^2 - 4C_k > 0, \ k \in \{i_{21}, i_{22}, \ldots, i_{2r}\} \subseteq \{i_1, i_2, \ldots, i_l\} \subseteq \{1, 2, \ldots, m\}, \tag{6.17}$$

$$\Delta_j = B_j^2 - 4C_j < 0, \ j \in \{i_{l+1}, i_{l+2}, \ldots, i_m\} \subseteq \{1, 2, \ldots, m\},$$

the $(2m + 1)$th-degree polynomial nonlinear system has $(2l + 1)$-equilibriums as

$$g(x^*) = b_1^{(i)} = -\frac{1}{2}B_i, \ g(x^*) = b_2^{(i)} = -\frac{1}{2}B_i$$

for $i \in \{i_{11}, i_{12}, \ldots, i_{ls}\}$,

$$g(x^*) = b_1^{(k)} = -\frac{1}{2}(B_k + \sqrt{\Delta_k}),$$

$$g(x^*) = b_2^{(k)} = -\frac{1}{2}(B_k - \sqrt{\Delta_k}) \tag{6.18}$$

for $k \in \{i_{21}, i_{22}, \ldots, i_{2r}\}$;

$$\{a_1, a_2, \ldots, a_{2l+1}\} = \text{sort}\{a, b_1^{(1)}, b_2^{(1)}, \ldots, b_1^{(l)}, b_2^{(l)}\}, \ a_\alpha < a_{\alpha+1},$$

$$S_\alpha = \{a_\alpha^{(s_\alpha)} | g(a_\alpha^{(s_\alpha)}) = a_\alpha, \ s_\alpha = 1, 2, \ldots, N_\alpha\} \cup \{\emptyset\},$$

$$\alpha = \{1, 2, \ldots, 2l + 1\}.$$

If

$$\{a_1, a_2, \ldots, a_{2l+1}\} = \text{sort}\,\{a, b_1^{(1)}, b_2^{(1)}, \ldots, b_1^{(l)}, b_2^{(l)}\},$$

$$a_{i_1} \equiv a_1 = \cdots = a_{l_1},$$

$$a_{i_2} \equiv a_{l_1+1} = \cdots = a_{l_1+l_2},$$

$$\vdots \tag{6.19}$$

$$a_{i_r} \equiv a_{\sum_{i=1}^{r-1} l_i + 1} = \cdots = a_{\sum_{i=1}^{r-1} l_i + l_r} = a_{2l+1}$$

$$\text{with } \sum_{s=1}^{r} l_s = 2l + 1,$$

then the corresponding standard functional form is

$$\dot{x} = a_0 \prod_{s=1}^{r} (x - a_{i_s})^{l_s} \prod_{k=l+1}^{m} \left[\left(x + \frac{1}{2} B_{i_k}\right)^2 + \frac{1}{4}(-\Delta_{i_k}) \right]. \tag{6.20}$$

The functional equilibrium separatrix flow is called an $(l_1\text{th XX}:l_2\text{th XX}:\ldots:l_r\text{th XX})$-flow.

(a) The functional equilibrium of $x^* = a_{i_p}^{(s)} \in S_{i_p}$ with $g(x^*) = a_{i_p}$ for $(l_p > 1)$-repeated functional equilibriums appearance or vanishing is called an l_pth XX bifurcation of functional equilibrium at a point $\mathbf{p} = \mathbf{p}_1 \in \partial\Omega_{12}$, and the functional bifurcation condition is

$$a_{i_p} \equiv a_{\sum_{i=1}^{p-1} l_i + 1} = \cdots = a_{\sum_{i=1}^{p-1} l_i + l_p} = -\frac{1}{2} B_{i_p},$$

$$\text{with } \Delta_{i_p} = B_{i_p}^2 - 4C_{i_p} = 0 \,(i_p \in \{i_1, i_2, \ldots i_l\}), \tag{6.21}$$

$$a_{\sum_{i=1}^{p-1} l_i + 1}^{+} \neq \cdots \neq a_{\sum_{i=1}^{p-1} l_i + l_p}^{+} \text{ or } a_{\sum_{i=1}^{p-1} l_i + 1}^{-} \neq \cdots \neq a_{\sum_{i=1}^{p-1} l_i + l_p}^{-}.$$

(b) The functional equilibrium of $x^* = a_{i_q}^{(s)} \in S_{i_q}$ with $g(x^*) = a_{i_q}$ for $(l_q > 1)$-repeated $(l_{q_1}\text{th XX}:l_{q_2}\text{th XX}:\ldots:l_{q_\beta}\text{th XX})$ functional equilibrium switching is called an l_qth XX bifurcation of functional equilibrium switching at a point $\mathbf{p} = \mathbf{p}_1 \in \partial\Omega_{12}$, and the switching bifurcation condition is

$$a_{i_q} \equiv a_{\sum_{i=1}^{q-1} l_i + 1} = \cdots = a_{\sum_{i=1}^{q-1} l_i + l_q},$$

$$a_{\sum_{i=1}^{q-1} l_i + 1}^{\pm} \neq \cdots \neq a_{\sum_{i=1}^{q-1} l_i + l_q}^{\pm};$$

$$l_q = \sum_{i=1}^{\beta} l_{q_i}; \; q = 1, 2, \ldots, r. \tag{6.22}$$

(c) The functional equilibrium of $x^* = a_{i_p}^{(s)} \in S_{i_p}$ with $g(x^*) = a_{i_p}$ for $(l_{p_1} \geq 1)$-repeated $(l_{p_{11}} \text{th XX} : l_{p_{12}} \text{th XX} : \ldots : l_{p_{1\beta}} \text{th XX})$ functional equilibrium appearance (or vanishing) and $(l_{p_2} \geq 2)$-repeated $(l_{p_{21}} \text{th XX} : l_{p_{22}} \text{th XX} : \ldots : l_{p_{2\beta}} \text{th XX})$ functional equilibriums switching is called an l_pth XX bifurcation of functional equilibrium at a point $\mathbf{p} = \mathbf{p}_1 \in \partial \Omega_{12}$, and the functional bifurcation condition is

$$a_{i_p} \equiv a_{\sum_{i=1}^{p-1} l_i + 1} = \cdots = a_{\sum_{i=1}^{p-1} l_i + l_p}$$
$$\text{with } \Delta_{i_p} = B_{i_p}^2 - 4C_{i_p} = 0 \, (i_p \in \{i_1, i_2, \ldots i_l\})$$
$$a_{\sum_{i=1}^{p-1} l_i + j_1}^+ \neq \cdots \neq a_{\sum_{i=1}^{p-1} l_i + j_{p_1}}^+ \quad \text{or } a_{\sum_{i=1}^{p_1-1} l_i + j_1}^- \neq \cdots \neq a_{\sum_{i=1}^{p_1-1} l_i + j_{p_1}}^-,$$
$$\text{for } \{j_1, j_2, \ldots, j_{p_1}\} \subseteq \{1, 2, \ldots, l_p\},$$
$$a_{\sum_{i=1}^{p-1} l_i + k_1}^{\pm} \neq \cdots \neq a_{\sum_{i=1}^{p-1} l_i + k_{p_2}}^{\pm}$$
$$\text{for } \{k_1, k_2, \ldots, k_{p_2}\} \subseteq \{1, 2, \ldots, l_p\},$$
$$\text{with } l_{p_1} + l_{p_2} = l_p; \, l_{p_2} = \sum_{i=1}^{\beta} l_{p_{2i}}, \, p = 1, 2, \ldots, r.$$

$$(6.23)$$

(iv) If
$$\Delta_i = B_i^2 - 4C_i > 0 \text{ for } i = 1, 2, \ldots, m, \quad (6.24)$$
the $(2m + 1)$th-degree polynomial nonlinear system has $(2m + 1)$ equilibriums as

$$g(x^*) = b_1^{(i)} = -\frac{1}{2}(B_i + \sqrt{\Delta_i}),$$
$$g(x^*) = b_2^{(i)} = -\frac{1}{2}(B_i - \sqrt{\Delta_i})$$
$$\text{for } i = 1, 2, \ldots, m; \quad (6.25)$$
$$\{a_1, a_2, \ldots, a_{2m+1}\} = \text{sort}\{a, b_1^{(1)}, b_2^{(1)}, \ldots, b_1^{(l)}, b_2^{(l)}\}, \, a_\alpha < a_{\alpha+1},$$
$$S_\alpha = \{a_\alpha^{(s_\alpha)} | g(a_\alpha^{(s_\alpha)}) = a_\alpha, \, s_\alpha = 1, 2, \ldots, N_\alpha\} \cup \{\emptyset\},$$
$$\alpha = \{1, 2, \ldots, 2m + 1\}.$$

(iv$_1$) If
$$b_r^{(i)} \neq b_s^{(j)} \text{ for } r, s \in \{1, 2\}; \, i, j = 1, 2, \ldots, m$$
$$\{a_1, a_2, \ldots, a_{2m}\} = \text{sort}\{a, b_1^{(1)}, b_2^{(1)}, \ldots, b_1^{(m)}, b_2^{(m)}\} \, (a_s < a_{s+1}),$$

$$(6.26)$$

then the corresponding standard functional form is

$$\dot{x} = a_0(g(x) - a_1)(g(x) - a_2)(g(x) - a_3)$$
$$\cdots (g(x) - a_{2m})(g(x) - a_{2m+1}).$$

$$(6.27)$$

This flow is formed with all simple equilibriums.

(a) If $a_0 dg/dx|_{x^*} > 0$, the separatrix flow with $(2m + 1)$-functional equilibriums is called a functional (SO:SI: ...:SO:SI:...:SI:SO) flow.

(b) If $a_0 dg/dx|_{x^*} < 0$, the separatrix flow with $(2m + 1)$-functional equilibriums is called a functional (SI:SO:...:SI:SO:...:SO:SI) flow.

(iv$_2$) If

$$\{a_1, a_2, \ldots, a_{2m+1}\} = \text{sort}\{a, b_1^{(1)}, b_2^{(1)}, \ldots, b_1^{(m)}, b_2^{(m)}\},$$

$$a_{i_1} \equiv a_1 = \cdots = a_{l_1},$$

$$a_{i_2} \equiv a_{l_1+1} = \cdots = a_{l_1+l_2},$$

$$\vdots \tag{6.28}$$

$$a_{i_r} \equiv a_{\sum_{i=1}^{r-1} l_i + 1} = \cdots = a_{\sum_{i=1}^{r-1} l_i + l_r} = a_{2m+1}$$

$$\text{with } \sum_{s=1}^{r} l_s = 2m + 1,$$

then, the corresponding standard functional form is

$$\dot{x} = a_0 \prod_{s=1}^{r} (g(x) - a_{i_s})^{l_s}. \tag{6.29}$$

The functional equilibrium separatrix flow is called an $(l_1 \text{th XX} : l_2 \text{th XX} : \ldots : l_r \text{th XX})$-flow. The functional equilibrium of $x^* = a_{i_p}^{(s)} \in S_{i_p}$ with $g(x^*) = a_{i_p}$ for l_p-repeated functional equilibriums switching is called a l_pth XX bifurcation of $(l_{p_1} \text{th XX} : l_{p_2} \text{th XX} : \ldots : l_{p_\beta} \text{th XX})$ functional equilibrium switching at a point $\mathbf{p} = \mathbf{p}_1 \in \partial\Omega_{12}$, and the functional switching bifurcation condition is

$$a_{i_p} \equiv a_{\sum_{i=1}^{p-1} l_i + 1} = \cdots = a_{\sum_{i=1}^{p-1} l_i + l_p},$$

$$a_{\sum_{i=1}^{p-1} l_i + 1}^{\pm} \neq \cdots \neq a_{\sum_{i=1}^{p-1} l_i + l_p}^{\pm};$$

$$l_p = \sum_{i=1}^{\beta} l_{p_i}, \quad p = 1, 2, \ldots, r. \tag{6.30}$$

Definition 6.2   Consider a $(2m + 1)$th-degree polynomial nonlinear functional dynamical system as

$$\begin{aligned}
\dot{x} &= A_0(\mathbf{p})(g(x))^{2m+1} + A_1(\mathbf{p})(g(x))^{2m} + \\
&\quad \cdots + A_{2m-1}(\mathbf{p})(g(x))^2 + A_{2m}g(x) + A_{2m+1}(\mathbf{p}) \\
&= a_0(\mathbf{p})(g(x) - a(\mathbf{p})) \prod_{i=1}^{n} [(g(x))^2 + B_i(\mathbf{p})g(x) + C_i(\mathbf{p})]^{q_i},
\end{aligned} \tag{6.31}$$

where $A_0(\mathbf{p}) \neq 0$ and

$$\mathbf{p} = (p_1, p_2, \ldots, p_m)^{\mathrm{T}}, \quad m = \sum_{i=1}^{n} q_i. \tag{6.32}$$

(i) If

$$\begin{aligned}
&\Delta_i = B_i^2 - 4C_i < 0 \text{ for } i = 1, 2, \ldots, n, \\
&S_1 = \{a_1^{(s)} | g(a_1^{(s_1)}) = a_1, \ s_1 = 1, 2, \ldots, N_1\} \cup \{\emptyset\}, \\
&a_1 = a,
\end{aligned} \tag{6.33}$$

the $(2m + 1)$th-degree polynomial nonlinear system has one equilibrium of $x^* = a_1^{(s)} \in S_1$ with $g(x^*) = a$, and the corresponding standard form is

$$\dot{x} = a_0(g(x) - a) \prod_{i=1}^{n} [(g(x) + \tfrac{1}{2} B_i)^2 + \tfrac{1}{4}(-\Delta_i)]^{q_i}. \tag{6.34}$$

The flow of such a functional system with one functional equilibrium is called a single functional equilibrium flow.

(a) If $a_0 dg/dx|_{x^*} > 0$, the functional equilibrium flow with $x^* = a_1^{(s)} \in S_1$ for $g(x^*) = a$ is called a functional source flow.

(b) If $a_0 dg/dx|_{x^*} < 0$, the functional equilibrium flow with $x^* = a_1^{(s)} \in S_1$ for $g(x^*) = a$ is called a functional sink flow.

(ii) If

$$\begin{aligned}
&\Delta_i = B_i^2 - 4C_i > 0, \ i \in \{i_1, i_2, \ldots, i_l\} \subseteq \{1, 2, \ldots, n\}, \\
&\Delta_j = B_j^2 - 4C_j < 0, \ j \in \{i_{l+1}, i_{l+2}, \ldots, i_n\} \subseteq \{1, 2, \ldots, n\},
\end{aligned} \tag{6.35}$$

the $(2m + 1)$th-degree polynomial nonlinear functional dynamical system has $(2l + 1)$-functional equilibriums as

$$g(x^*) = b_1^{(i)} = -\frac{1}{2}(B_i + \sqrt{\Delta_i}),$$

$$g(x^*) = b_2^{(i)} = -\frac{1}{2}(B_i - \sqrt{\Delta_i}),$$

$$i \in \{i_1, i_2, \ldots, i_l\} \subseteq \{1, 2, \ldots, n\};$$

$$\{a_1, a_2, \ldots, a_{2l+1}\} = \text{sort}\{a, b_1^{(1)}, b_2^{(1)}, \ldots, b_1^{(l)}, b_2^{(l)}\}, \ a_\alpha < a_{\alpha+1},$$

$$S_\alpha = \{a_\alpha^{(s_\alpha)}| g(a_\alpha^{(s_\alpha)}) = a_\alpha, \ s_\alpha = 1, 2, \ldots, N_\alpha\} \cup \{\emptyset\},$$

$$\alpha = \{1, 2, \ldots, 2l + 1\}. \tag{6.36}$$

(ii$_1$) If

$$b_r^{(i)} \neq b_s^{(j)} \text{ for } r, s \in \{1, 2\}; \ i, j = 1, 2, \ldots, l$$

$$\{a_1, a_2, \ldots, a_{2l+1}\} = \text{sort}\{a, \underbrace{b_1^{(1)}, b_2^{(1)}}_{q_1 \text{ sets}}, \ldots, \underbrace{b_1^{(r)}, b_2^{(r)}}_{q_r \text{ sets}}\}, \ a_s \leq a_{s+1}, \tag{6.37}$$

then the corresponding standard form is

$$\dot{x} = a_0 \prod_{s=1}^{2l+1} (g(x) - a_s)^{l_s} \prod_{k=l+1}^{n} [(g(x) + \frac{1}{2}B_{i_k})^2 + \frac{1}{4}(-\Delta_{i_k})]^{q_{i_k}} \tag{6.38}$$

with $l_s \in \{q_{i_1}, q_{i_2}, \ldots, q_{i_l}, 1\}$.

The equilibrium separatrix flow is called a $(l_1\text{th XX} : l_2\text{th XX} : \ldots : l_{2l+1}\text{th XX})$-flow.

(a$_1$) For $a_0 dg/dx|_{x^*} > 0 \ (p = 1, 2, \ldots, 2l + 1)$,

$$l_p\text{th XX} = \begin{cases} (2r_p - 1)^{\text{th}} \text{ order source, for } \alpha_p = 2M_p - 1, \ l_p = 2r_p - 1; \\ (2r_p - 1)^{\text{th}} \text{ order sink, for } \alpha_p = 2M_p, \ l_p = 2r_p - 1, \end{cases} \tag{6.39}$$

where

$$\alpha_p = \sum_{s=p}^{2l+1} l_s. \tag{6.40}$$

(a$_2$) For $a_0 dg/dx|_{x^*} < 0 \ (p = 1, 2, \ldots, 2l + 1)$,

$$l_p\text{th XX} = \begin{cases} (2r_p - 1)^{\text{th}} \text{ order sink, for } \alpha_p = 2M_p - 1, \ l_p = 2r_p - 1; \\ (2r_p - 1)^{\text{th}} \text{ order source, for } \alpha_p = 2M_p, \ l_p = 2r_p - 1. \end{cases} \tag{6.41}$$

(a₃) For $a_0 > 0$ and $dg/dx|_{x^*} \neq 0$ $(p = 1, 2, \ldots, 2l + 1)$,

$$l_p \text{th XX} = \begin{cases} (2r_p)^{\text{th}} \text{ order lower-saddle, for } \alpha_p = 2M_p - 1, \ l_p = 2r_p; \\ (2r_p)^{\text{th}} \text{ order upper-saddle, for } \alpha_p = 2M_p, \ l_p = 2r_p. \end{cases}$$

(6.42)

(a₄) For $a_0 < 0$ and $dg/dx|_{x^*} \neq 0$ $(p = 1, 2, \ldots, 2l + 1)$,

$$l_p \text{th XX} = \begin{cases} (2r_p)^{\text{th}} \text{ order upper-saddle, for } \alpha_p = 2M_p - 1, \ l_p = 2r_p; \\ (2r_p)^{\text{th}} \text{ order lower-saddle, for } \alpha_p = 2M_p, \ l_p = 2r_p. \end{cases}$$

(6.43)

(ii₂) If

$$\{a_1, a_2, \ldots, a_{2l+1}\} = \text{sort}\{a, \underbrace{b_1^{(1)}, b_2^{(1)}}_{q_1 \text{ sets}}, \ldots, \underbrace{b_1^{(r)}, b_2^{(r)}}_{q_r \text{ sets}}\},$$

$$a_{i_1} \equiv a_1 = \cdots = a_{l_1},$$
$$a_{i_2} \equiv a_{l_1+1} = \cdots = a_{l_1+l_2},$$
$$\vdots$$
$$a_{i_r} \equiv a_{\sum_{i=1}^{r-1} l_i + 1} = \cdots = a_{\sum_{i=1}^{r-1} l_i + l_r} = a_{2l+1}$$

$$\text{with } \sum_{s=1}^{r} l_s = 2l + 1,$$

(6.44)

then, the corresponding standard functional form is

$$\dot{x} = a_0 \prod_{s=1}^{r} (g(x) - a_{i_s})^{l_s} \prod_{k=l+1}^{n} [(g(x) + \frac{1}{2} B_{i_k})^2 + \frac{1}{4}(-\Delta_{i_k})]^{q_{i_k}}.$$

(6.45)

The functional equilibrium separatrix flow is called an $(l_1 \text{th XX} : l_2 \text{th XX} : \ldots : l_r \text{th XX})$-functional flow.

(a₁) For $a_0 dg/dx|_{x^*} > 0$ $(p = 1, 2, \ldots, r)$,

$$l_p \text{th XX} = \begin{cases} (2r_p - 1)^{\text{th}} \text{ order source, for } \alpha_p = 2M_p - 1, \ l_p = 2r_p - 1; \\ (2r_p - 1)^{\text{th}} \text{ order sink, for } \alpha_p = 2M_p, \ l_p = 2r_p - 1, \end{cases}$$

(6.46)

where

$$\alpha_p = \sum_{s=p}^{r} l_s.$$

(6.47)

(a_2) For $a_0 dg/dx|_{x*} < 0$ $(p = 1, 2, \ldots, r)$,

$$l_p \text{th XX} = \begin{cases} (2r_p - 1)^{\text{th}} \text{ order sink, for } \alpha_p = 2M_p - 1, \, l_p = 2r_p - 1; \\ (2r_p - 1)^{\text{th}} \text{ order source, for } \alpha_p = 2M_p, \, l_p = 2r_p - 1. \end{cases}$$

(6.48)

(a_3) For $a_0 > 0$ and $dg/dx|_{x*} \neq 0$ $(p = 1, 2, \ldots, r)$,

$$l_p \text{th XX} = \begin{cases} (2r_p)^{\text{th}} \text{ order lower-saddle, for } \alpha_p = 2M_p - 1, \, l_p = 2r_p; \\ (2r_p)^{\text{th}} \text{ order upper-saddle, for } \alpha_p = 2M_p, \, l_p = 2r_p. \end{cases}$$

(6.49)

(a_4) For $a_0 < 0$ and $dg/dx|_{x*} \neq 0$ $(p = 1, 2, \ldots, r)$,

$$l_p \text{th XX} = \begin{cases} (2r_p)^{\text{th}} \text{ order upper-saddle, for } \alpha_p = 2M_p - 1, \, l_p = 2r_p; \\ (2r_p)^{\text{th}} \text{ order lower-saddle, for } \alpha_p = 2M_p, \, l_p = 2r_p. \end{cases}$$

(6.50)

(b) The functional equilibrium of $x^* = a_{i_p}^{(s)} \in S_{i_p}$ with $g(x^*) = a_{i_p}$ for $(l_p > 1)$-repeated functional equilibriums switching is called an l_pth XX bifurcation of $(l_{p_1}$th XX $: l_{p_2}$th XX $: \ldots : l_{p_\beta}$th XX$)$ functional equilibrium switching at a point $\mathbf{p} = \mathbf{p}_1 \in \partial\Omega_{12}$, and the functional switching bifurcation condition is

$$a_{i_p} \equiv a_{\sum_{i=1}^{p-1} l_i + 1} = \cdots = a_{\sum_{i=1}^{p-1} l_i + l_p},$$
$$a_{\sum_{i=1}^{p-1} l_i + 1}^{\pm} \neq \cdots \neq a_{\sum_{i=1}^{p-1} l_i + l_p}^{\pm};$$
$$l_p = \sum_{i=1}^{\beta} l_{p_i}, \quad p = 1, 2, \ldots, r.$$

(6.51)

(iii) If

$$\Delta_i = B_i^2 - 4C_i = 0, \, i \in \{i_{11}, i_{12}, \ldots, i_{1_s}\} \subseteq \{i_1, i_2, \ldots, i_l\} \subseteq \{1, 2, \ldots, n\},$$
$$\Delta_k = B_k^2 - 4C_k > 0, \, k \in \{i_{21}, i_{22}, \ldots, i_{2r}\} \subseteq \{i_1, i_2, \ldots, i_l\} \subseteq \{1, 2, \ldots, n\},$$
$$\Delta_j = B_j^2 - 4C_j < 0, \, j \in \{i_{l+1}, i_{l+2}, \ldots, i_n\} \subseteq \{1, 2, \ldots, n\} \text{ with } i \neq j \neq k,$$

(6.52)

the $(2m + 1)$th-degree polynomial nonlinear functional dynamical system has $(2l + 1)$-functional equilibriums as

$$g(x^*) = b_1^{(i)} = -\frac{1}{2}B_i, \ g(x^*) = b_2^{(i)} = -\frac{1}{2}B_i$$

for $i \in \{i_{11}, i_{12}, \ldots, i_{ls}\}$,

$$g(x^*) = b_1^{(k)} = -\frac{1}{2}(B_k + \sqrt{\Delta_k}),$$

$$g(x^*) = b_2^{(k)} = -\frac{1}{2}(B_k - \sqrt{\Delta_k}) \tag{6.53}$$

for $k \in \{i_{21}, i_{22}, \ldots, i_{2r}\}$,

$$\{a_1, a_2, \ldots, a_{2l+1}\} = \text{sort}\{a, b_1^{(1)}, b_2^{(1)}, \ldots, b_1^{(l)}, b_2^{(l)}\}, \ a_\alpha < a_{\alpha+1},$$

$$S_\alpha = \{a_\alpha^{(s_\alpha)} | g(a_\alpha^{(s_\alpha)}) = a_\alpha, \ s_\alpha = 1, 2, \ldots, N_\alpha\} \cup \{\emptyset\},$$

$$\alpha = \{1, 2, \ldots, r\}.$$

If

$$\{a_1, a_2, \ldots, a_{2l+1}\} = \text{sort}\{a, b_1^{(1)}, b_2^{(1)}, \ldots, b_1^{(l)}, b_2^{(l)}\},$$

$$a_{i_1} \equiv a_1 = \cdots = a_{l_1},$$

$$a_{i_2} \equiv a_{l_1+1} = \cdots = a_{l_1+l_2},$$

$$\vdots \tag{6.54}$$

$$a_{i_r} \equiv a_{\sum_{i=1}^{r-1} l_i + 1} = \cdots = a_{\sum_{i=1}^{r-1} l_i + l_r} = a_{2l+1}$$

$$\text{with} \ \sum_{s=1}^{r} l_s = 2l + 1,$$

then the corresponding standard functional form is

$$\dot{x} = a_0 \prod_{s=1}^{r} (g(x) - a_{i_s})^{l_s} \prod_{k=l+1}^{n} [(g(x) + \frac{1}{2}B_{i_k})^2 + \frac{1}{4}(-\Delta_{i_k})]^{q_{i_k}}. \tag{6.55}$$

The functional equilibrium separatrix flow is called an $(l_1$th XX$: l_2$th XX$: \ldots : l_r$th XX$)$-functional flow.

(a) The functional equilibrium of $x^* = a_{i_p}^{(s)} \in S_{i_p}$ with $g(x^*) = a_{i_p}$ for $(l_p > 1)$-repeated functional equilibriums appearance or vanishing is called an l_pth XX bi-furcation of functional equilibrium at a point $\mathbf{p} = \mathbf{p}_1 \in \partial\Omega_{12}$, and the functional

bifurcation condition is

$$a_{i_p} \equiv a_{\sum_{i=1}^{p-1} l_i + 1} = \cdots = a_{\sum_{i=1}^{p-1} l_i + l_p} = -\frac{1}{2} B_{i_p},$$

$$\text{with } \Delta_{i_p} = B_{i_p}^2 - 4C_{i_p} = 0 \, (i_p \in \{i_1, i_2, \ldots i_l\}), \tag{6.56}$$

$$a_{\sum_{i=1}^{p-1} l_i + 1}^+ \neq \cdots \neq a_{\sum_{i=1}^{p-1} l_i + l_p}^+ \quad \text{or} \quad a_{\sum_{i=1}^{p-1} l_i + 1}^- \neq \cdots \neq a_{\sum_{i=1}^{p-1} l_i + l_p}^-.$$

(b) The functional equilibrium of $x^* = a_{i_q}^{(s)} \in S_{i_q}$ with $g(x^*) = a_{i_q}$ for $(l_q > 1)$-repeated functional equilibriums switching is called an l_qth XX bifurcation of $(l_{q_1}$th XX $: l_{q_2}$th XX $: \ldots : l_{q_\beta}$th XX$)$ equilibrium switching at a point $\mathbf{p} = \mathbf{p}_1 \in \partial\Omega_{12}$, and the bifurcation condition is

$$a_{i_q} \equiv a_{\sum_{i=1}^{q-1} l_i + 1} = \cdots = a_{\sum_{i=1}^{q-1} l_i + l_q},$$

$$a_{\sum_{i=1}^{q-1} l_i + 1}^\pm \neq \cdots \neq a_{\sum_{i=1}^{q-1} l_i + l_q}^\pm; \tag{6.57}$$

$$l_q = \sum_{i=1}^{\beta} l_{q_i}, \quad q = 1, 2, \ldots, r.$$

(iv) If

$$\Delta_i = B_i^2 - 4C_i > 0 \text{ for } i = 1, 2, \ldots, n, \tag{6.58}$$

the $(2m + 1)$th-degree polynomial nonlinear functional dynamical system has $(2n + 1)$-functional equilibriums as

$$g(x^*) = b_1^{(i)} = -\frac{1}{2}(B_i + \sqrt{\Delta_i}),$$

$$g(x^*) = b_2^{(i)} = -\frac{1}{2}(B_i - \sqrt{\Delta_i})$$

$$\text{for } i = 1, 2, \ldots, n; \tag{6.59}$$

$$\{a_1, a_2, \ldots, a_{2n+1}\} = \text{sort} \{a, b_1^{(1)}, b_2^{(1)}, \ldots, b_1^{(n)}, b_2^{(n)}\}, \quad a_\alpha < a_{\alpha+1},$$

$$S_\alpha = \{a_\alpha^{(s_\alpha)} | g(a_\alpha^{(s_\alpha)}) = a_\alpha, \; s_\alpha = 1, 2, \ldots, N_\alpha\} \cup \{\emptyset\},$$

$$\alpha = \{1, 2, \ldots, 2n + 1\}.$$

(iv$_1$) If

$$b_r^{(i)} \neq b_s^{(j)} \text{ for } r, s \in \{1, 2\}, \; (i, j = 1, 2, \ldots, n);$$

$$\{a_1, a_2, \ldots, a_{2n+1}\} = \text{sort} \{a, \underbrace{b_1^{(1)}, b_2^{(1)}}_{q_1 \text{ sets}}, \ldots, \underbrace{b_1^{(n)}, b_2^{(n)}}_{q_n \text{ sets}}\} \; (a_s \leq a_{s+1}), \tag{6.60}$$

then the corresponding standard form is

$$\dot{x} = a_0 \prod_{s=1}^{2n+1} (x - a_s)^{l_s} \text{ with } l_s \in \{q_{i_1}, q_{i_2}, \ldots, q_{i_n}, 1\}. \qquad (6.61)$$

The functional equilibrium separatrix flow is called an $(l_1 \text{th XX} : l_2 \text{th XX} : \ldots : l_{2n+1} \text{th XX})$-functional flow.

(a_1) For $a_0 dg/dx|_{x^*} > 0 \ (p = 1, 2, \ldots, 2n + 1)$,

$$l_p \text{th XX} = \begin{cases} (2r_p - 1)^{\text{th}} \text{ order source, for } \alpha_p = 2M_p - 1, \ l_p = 2r_p - 1; \\ (2r_p - 1)^{\text{th}} \text{ order sink, for } \alpha_p = 2M_p, \ l_p = 2r_p - 1, \end{cases}$$
$$(6.62)$$

where

$$\alpha_p = \sum_{s=p}^{2n+1} l_s. \qquad (6.63)$$

(a_2) For $a_0 dg/dx|_{x^*} < 0 \ (p = 1, 2, \ldots, 2n + 1)$,

$$l_p \text{th XX} = \begin{cases} (2r_p - 1)^{\text{th}} \text{ order sink, for } \alpha_p = 2M_p - 1, \ l_p = 2r_p - 1; \\ (2r_p - 1)^{\text{th}} \text{ order source, for } \alpha_p = 2M_p, \ l_p = 2r_p - 1. \end{cases}$$
$$(6.64)$$

(a_3) For $a_0 > 0$ and $dg/dx|_{x^*} \neq 0 \ (p = 1, 2, \ldots, 2n + 1)$,

$$l_p \text{th XX} = \begin{cases} (2r_p)^{\text{th}} \text{ order lower-saddle, for } \alpha_p = 2M_p - 1, \ l_p = 2r_p; \\ (2r_p)^{\text{th}} \text{ order upper-saddle, for } \alpha_p = 2M_p, \ l_p = 2r_p. \end{cases}$$
$$(6.65)$$

(a_4) For $a_0 < 0$ and $dg/dx|_{x^*} \neq 0 \ (p = 1, 2, \ldots, 2n + 1)$,

$$l_p \text{th XX} = \begin{cases} (2r_p)^{\text{th}} \text{ order upper-saddle, for } \alpha_p = 2M_p - 1, \ l_p = 2r_p; \\ (2r_p)^{\text{th}} \text{ order lower-saddle, for } \alpha_p = 2M_p, \ l_p = 2r_p. \end{cases}$$
$$(6.66)$$

(iv$_2$) If

$$\{a_1, a_2, \ldots, a_{2n+1}\} = \text{sort}\{a, \underbrace{b_1^{(1)}, b_2^{(1)}}_{q_1 \text{ sets}}, \ldots, \underbrace{b_1^{(n)}, b_2^{(n)}}_{q_n \text{ sets}}\},$$

$$a_{i_1} \equiv a_1 = \cdots = a_{l_1},$$

$$a_{i_2} \equiv a_{l_1+1} = \cdots = a_{l_1+l_2},$$

$$\vdots \qquad\qquad\qquad\qquad\qquad\qquad (6.67)$$

$$a_{i_r} \equiv a_{\sum_{i=1}^{r-1} l_i + 1} = \cdots = a_{\sum_{i=1}^{r-1} l_i + l_r} = a_{2n+1},$$

$$\text{with } \sum_{s=1}^{r} l_s = 2n + 1,$$

then the corresponding standard functional form is

$$\dot{x} = a_0 \prod_{s=1}^{r} (x - a_{i_s})^{l_s}. \qquad (6.68)$$

The functional equilibrium separatrix flow is called an $(l_1\text{th XX} : l_2\text{th XX} : \ldots :$ $l_r\text{th XX})$-flow. The functional equilibrium of $x^* = a_{i_p}^{(s)} \in S_{i_p}$ with $g(x^*) = a_{i_p}$ for l_p-repeated functional equilibriums switching is called an $l_p\text{th XX}$ switching bifurcation of $(l_{p_1}\text{th XX} : l_{p_2}\text{th XX} : \ldots : l_{p_\beta}\text{th XX})$-functional equilibrium at a point $\mathbf{p} = \mathbf{p}_1 \in \partial\Omega_{12}$, and the functional switching bifurcation condition is

$$a_{i_p} \equiv a_{\sum_{i=1}^{p-1} l_i + 1} = \cdots = a_{\sum_{i=1}^{p-1} l_i + l_p},$$

$$a_{\sum_{i=1}^{p-1} l_i + 1}^{\pm} \neq \cdots \neq a_{\sum_{i=1}^{p-1} l_i + l_p}^{\pm}; \qquad (6.69)$$

$$l_p = \sum_{i=1}^{\beta} l_{p_i}, \quad p = 1, 2, \ldots, r.$$

Definition 6.3   Consider a 1-dimensional, $(2m + 1)$th-degree polynomial nonlinear functional dynamical system

$$\dot{x} = A_0(\mathbf{p})(g(x))x^{2m+1} + A_1(\mathbf{p})(g(x))^{2m} +$$

$$\cdots + A_{2m-1}(\mathbf{p})(g(x))^2 + A_{2m}g(x) + A_{2m+1}(\mathbf{p})$$

$$= a_0(\mathbf{p}) \prod_{s=1}^{r} (g(x) - c_{i_s}(\mathbf{p}))^{l_s} \prod_{i=r+1}^{n} [(g(x))^2 + B_i(\mathbf{p})g(x) + C_i(\mathbf{p})]^{q_i}, \qquad (6.70)$$

where $A_0(\mathbf{p}) \neq 0$, and

$$\sum_{s=1}^{r} l_s = 2l + 1, \quad \sum_{i=r+1}^{n} q_i = (m-l), \quad \mathbf{p} = (p_1, p_2, \ldots, p_m)^{\mathrm{T}}. \tag{6.71}$$

(i) If

$$
\begin{aligned}
\Delta_i &= B_i^2 - 4C_i < 0 \text{ for } i = r+1, r+2, \ldots, n, \\
\{a_1, a_2, \ldots, a_r\} &= \text{sort}\{c_1, c_2, \ldots, c_r\}, \text{ with } a_\alpha < a_{\alpha+1}, \\
S_\alpha &= \{a_\alpha^{(s_\alpha)} | g(a_\alpha^{(s_\alpha)}) = a_\alpha, \ s_\alpha = 1, 2, \ldots, N_\alpha\} \cup \{\emptyset\}, \\
\alpha &= \{1, 2, \ldots, r\},
\end{aligned} \tag{6.72}
$$

the $(2m+1)$th-degree polynomial functional dynamical system has functional equilibriums of $x^* = a_{i_p}^{(s)} \in S_{i_p}$ with $g(x^*) = a_{i_p}$ $(p = 1, 2, \ldots, r)$, and the corresponding standard functional form is

$$\dot{x} = a_0(\mathbf{p}) \prod_{j=1}^{r} (g(x) - a_{i_j})^{l_j} \prod_{i=r+1}^{n} [(g(x) + \tfrac{1}{2} B_i)^2 + \tfrac{1}{4}(-\Delta_i)]^{l_i}. \tag{6.73}$$

The functional equilibrium separatrix flow is called an $(l_1 \text{th XX}: l_2 \text{th XX}: \ldots : l_r \text{th XX})$-flow.

(a$_1$) For $a_0 dg/dx|_{x^*} > 0$ $(p = 1, 2, \ldots, r)$,

$$l_p \text{th XX} = \begin{cases} (2r_p - 1)^{\text{th}} \text{ order source, for } \alpha_p = 2M_p - 1, \ l_p = 2r_p - 1; \\ (2r_p - 1)^{\text{th}} \text{ order sink, for } \alpha_p = 2M_p, \ l_p = 2r_p - 1, \end{cases} \tag{6.74}$$

where

$$\alpha_p = \sum_{s=p}^{r} l_s. \tag{6.75}$$

(a$_2$) For $a_0 dg/dx|_{x^*} < 0$ $(p = 1, 2, \ldots, r)$,

$$l_p \text{th XX} = \begin{cases} (2r_p - 1)^{\text{th}} \text{ order sink, for } \alpha_p = 2M_p - 1, \ l_p = 2r_p - 1; \\ (2r_p - 1)^{\text{th}} \text{ order source, for } \alpha_p = 2M_p, \ l_p = 2r_p - 1. \end{cases} \tag{6.76}$$

(a$_3$) For $a_0 > 0$ and $dg/dx|_{x^*} \neq 0$ $(p = 1, 2, \ldots, r)$,

$$l_p \text{th XX} = \begin{cases} (2r_p)^{\text{th}} \text{ order lower-saddle, for } \alpha_p = 2M_p - 1, \ l_p = 2r_p; \\ (2r_p)^{\text{th}} \text{ order upper-saddle, for } \alpha_p = 2M_p, \ l_p = 2r_p. \end{cases} \tag{6.77}$$

(a$_4$) For $a_0 < 0$ and $dg/dx|_{x*} \neq 0$ ($p = 1, 2, \ldots, r$),

$$l_p \text{th XX} = \begin{cases} (2r_p)^{\text{th}} \text{ order upper-saddle, for } \alpha_p = 2M_p - 1, \ l_p = 2r_p; \\ (2r_p)^{\text{th}} \text{ order lower-saddle, for } \alpha_p = 2M_p, \ l_p = 2r_p. \end{cases}$$

(6.78)

(ii) If

$$\Delta_i = B_i^2 - 4C_i > 0, \ i = j_1, j_2, \ldots, j_s \in \{l+1, l+2, \ldots, n\},$$
$$\Delta_j = B_j^2 - 4C_j < 0, \ j = j_{s+1}, j_{s+2}, \ldots, j_n \in \{l+1, l+2, \ldots, n\}$$

(6.79)

$$\text{with } s \in \{1, \ldots, n-l\},$$

the $(2m + 1)$th-degree polynomial nonlinear functional dynamical system has $2n_2$-functional equilibriums as

$$g(x^*) = b_1^{(i)} = -\frac{1}{2}(B_i + \sqrt{\Delta_i}),$$
$$g(x^*) = b_2^{(i)} = -\frac{1}{2}(B_i - \sqrt{\Delta_i}),$$
$$i \in \{j_1, j_2, \ldots, j_{n_1}\} \subseteq \{l+1, l+2, \ldots, n\};$$
$$\{a_1, a_2, \ldots, a_{2n_2+1}\} = \text{sort}\{c_1, c_2, \ldots, c_{2l+1}, \underbrace{b_1^{(r+1)}, b_2^{(r+1)}}_{q_{r+1}\text{-sets}}, \ldots, \underbrace{b_1^{(n_1)}, b_2^{(n_1)}}_{q_{n_1}\text{-sets}}\};$$

$$\text{with } a_\alpha < a_{\alpha+1}$$
$$S_\alpha = \{a_\alpha^{(s_\alpha)} | g(a_\alpha^{(s_\alpha)}) = a_\alpha, \ s_\alpha = 1, 2, \ldots, N_\alpha\} \cup \{\emptyset\},$$
$$\alpha = \{1, 2, \ldots, 2n_2 + 1\}.$$

(6.80)

If

$$\{a_1, a_2, \ldots, a_{2n_2+1}\} = \text{sort}\{c_1, c_2, \ldots, c_{2l+1}, \underbrace{b_1^{(r+1)}, b_2^{(r+1)}}_{q_{r+1}\text{ sets}}, \ldots, \underbrace{b_1^{(n_1)}, b_2^{(n_1)}}_{q_{n_1}\text{ sets}}\},$$

$$a_{i_1} \equiv a_1 = \cdots = a_{l_1},$$
$$a_{i_2} \equiv a_{l_1+1} = \cdots = a_{l_1+l_2},$$
$$\vdots$$
$$a_{i_{n_1}} \equiv a_{\sum_{i=1}^{n_1-1} l_i+1} = \cdots = a_{\sum_{i=1}^{n_1-1} l_i+l_{n_1}} = a_{2n_2+1}$$

$$\text{with } \sum_{s=1}^{n_1} l_s = 2n_2 + 1,$$

(6.81)

then the corresponding standard functional form is

$$\dot{x} = a_0 \prod_{s=1}^{n_1} (g(x) - a_{i_s})^{l_s} \prod_{i=n_2+1}^{n} [(g(x) + \frac{1}{2}B_i)^2 + \frac{1}{4}(-\Delta_i)]^{q_i}. \tag{6.82}$$

The functional equilibrium separatrix flow is called an $(l_1 \text{th XX} : l_2 \text{th XX} : \ldots : l_{n_1} \text{th XX})$-functional flow.

(a$_1$) For $a_0 dg/dx|_{x^*} > 0$ $(p = 1, 2, \ldots, r, r+1, \ldots, n_1)$,

$$l_p \text{th XX} = \begin{cases} (2r_p - 1)^{\text{th}} \text{ order source, for } \alpha_p = 2M_p - 1, \ l_p = 2r_p - 1; \\ (2r_p - 1)^{\text{th}} \text{ order sink, for } \alpha_p = 2M_p, \ l_p = 2r_p - 1, \end{cases}$$

$$\tag{6.83}$$

where

$$\alpha_p = \sum_{s=p}^{n_1} l_s. \tag{6.84}$$

(a$_2$) For $a_0 dg/dx|_{x^*} < 0$ $(p = 1, 2, \ldots, r, r+1, \ldots, n_1)$,

$$l_p \text{th XX} = \begin{cases} (2r_p - 1)^{\text{th}} \text{ order sink, for } \alpha_p = 2M_p - 1, \ l_p = 2r_p - 1; \\ (2r_p - 1)^{\text{th}} \text{ order source, for } \alpha_p = 2M_p, \ l_p = 2r_p - 1. \end{cases}$$

$$\tag{6.85}$$

(a$_3$) For $a_0 > 0$ and $dg/dx|_{x^*} \neq 0$ $(p = 1, 2, \ldots, r, r+1, \ldots, n_1)$,

$$l_p \text{th XX} = \begin{cases} (2r_p)^{\text{th}} \text{ order lower-saddle, for } \alpha_p = 2M_p - 1, \ l_p = 2r_p; \\ (2r_p)^{\text{th}} \text{ order upper-saddle, for } \alpha_p = 2M_p, \ l_p = 2r_p. \end{cases}$$

$$\tag{6.86}$$

(a$_4$) For $a_0 < 0$ and $dg/dx|_{x^*} \neq 0$ $(p = 1, 2, \ldots, r, r+1, \ldots, n_1)$,

$$l_p \text{th XX} = \begin{cases} (2r_p)^{\text{th}} \text{ order upper-saddle, for } \alpha_p = 2M_p - 1, \ l_p = 2r_p; \\ (2r_p)^{\text{th}} \text{ order lower-saddle, for } \alpha_p = 2M_p, \ l_p = 2r_p. \end{cases}$$

$$\tag{6.87}$$

(b) The functional equilibrium of $x^* = a_{i_p}^{(s)} \in S_{i_p}$ with $g(x^*) = a_{i_p}$ for $(l_p > 1)$-repeated functional equilibriums switching is called an l_pth XX switching bifurcation of $(l_{p_1} \text{th XX} : l_{p_2} \text{th XX} : \ldots : l_{p_\beta} \text{th XX})$ functional equilibrium at a point $\mathbf{p} = \mathbf{p}_1 \in \partial \Omega_{12}$, and the functional bifurcation condition is

$$a_{i_p} \equiv a_{\sum_{i=1}^{p-1} l_i + 1} = \cdots = a_{\sum_{i=1}^{p-1} l_i + l_p},$$

$$a_{\sum_{i=1}^{p-1} l_i + 1}^{\pm} \neq \cdots \neq a_{\sum_{i=1}^{p-1} l_i + l_p}^{\pm};$$

$$l_p = \sum_{i=1}^{\beta} l_{p_i}, \ p = 1, 2, \ldots, n_1. \tag{6.88}$$

(iii) If

$$\Delta_i = B_i^2 - 4C_i = 0,$$
$$\text{for } i \in \{i_{11}, i_{12}, \ldots, i_{1_s}\} \subseteq \{i_{l+1}, i_{l+2}, \ldots, i_{n_2}\} \subseteq \{l+1, l+2, \ldots, n\},$$
$$\Delta_k = B_k^2 - 4C_k > 0,$$
$$\text{for } k \in \{i_{21}, i_{22}, \ldots, i_{2r}\} \subseteq \{i_{l+1}, i_{l+2}, \ldots, i_{n_2}\} \subseteq \{l+1, l+2, \ldots, n\},$$
$$\Delta_j = B_j^2 - 4C_j < 0,$$
$$\text{for } j \in \{i_{n_2+1}, i_{n_2+2}, \ldots, i_n\} \subseteq \{l+1, l+2, \ldots, n\},$$

$$(6.89)$$

the $(2m+1)$th-degree polynomial nonlinear system has $(2n_2+1)$-equilibriums as

$$g(x^*) = b_1^{(i)} = -\frac{1}{2}B_i, \ \ g(x^*) = b_2^{(i)} = -\frac{1}{2}B_i,$$
$$\text{for } i \in \{i_{11}, i_{12}, \ldots, i_{1_s}\};$$
$$g(x^*) = b_1^{(k)} = -\frac{1}{2}(B_k + \sqrt{\Delta_k}),$$
$$g(x^*) = b_2^{(k)} = -\frac{1}{2}(B_k - \sqrt{\Delta_k})$$
$$\text{for } i \in \{i_{21}, i_{22}, \ldots, i_{2r}\};$$

$$(6.90)$$

$$\{a_1, a_2, \ldots, a_{2n_2+1}\} = \text{sort}\{a, c_1, c_2, \ldots, c_{2l}, \underbrace{b_1^{(r)}, b_2^{(r)}}_{q_r \text{ sets}}, \ldots, \underbrace{b_1^{(n_1)}, b_2^{(n_1)}}_{q_{n_1} \text{ sets}}\},$$

$$\text{with } a_\alpha < a_{\alpha+1}$$
$$S_\alpha = \{a_\alpha^{(s_\alpha)} | g(a_\alpha^{(s_\alpha)}) = a_\alpha, \ s_\alpha = 1, 2, \ldots, N_\alpha\} \cup \{\emptyset\},$$
$$\alpha = \{1, 2, \ldots, 2n_2 + 1\}.$$

If

$$\{a_1, a_2, \ldots, a_{2n_2+1}\} = \text{sort}\{a, c_1, c_2, \ldots, c_{2l}, \underbrace{b_1^{(r)}, b_2^{(r)}}_{q_r \text{ sets}}, \ldots, \underbrace{b_1^{(n_1)}, b_2^{(n_1)}}_{q_{n_1} \text{ sets}}\},$$

$$a_{i_1} \equiv a_1 = \cdots = a_{l_1},$$
$$a_{i_2} \equiv a_{l_1+1} = \cdots = a_{l_1+l_2},$$
$$\vdots$$
$$a_{i_{n_1}} \equiv a_{\sum_{i=1}^{n_1-1} l_i + 1} = \cdots = a_{\sum_{i=1}^{n_1-1} l_i + l_{n_1}} = a_{2n_2+1}$$

$$(6.91)$$

$$\text{with } \sum_{s=1}^{n_1} l_s = 2n_2 + 1,$$

then the corresponding standard form is

$$\dot{x} = a_0 \prod_{s=1}^{n_1} (g(x) - a_{i_s})^{l_s} \prod_{i=n_2+1}^{n} [(g(x) + \frac{1}{2}B_i)^2 + \frac{1}{4}(-\Delta_i)]^{q_i}. \tag{6.92}$$

The functional equilibrium separatrix flow is called an $(l_1\text{th XX}:l_2\text{th XX}:\dots:l_{n_1}\text{th XX})$-functional flow.

(a) The functional equilibrium of $x^* = a_{i_p}^{(s)} \in S_{i_p}$ with $g(x^*) = a_{i_p}$ for $(l_p > 1)$-repeated functional equilibriums appearance or vanishing is called an l_pth XX bifurcation of functional equilibrium at a point $\mathbf{p} = \mathbf{p}_1 \in \partial\Omega_{12}$, and the functional bifurcation condition is

$$a_{i_p} \equiv a_{\sum_{i=1}^{p-1} l_i+1} = \dots = a_{\sum_{i=1}^{p-1} l_i+l_p} = -\frac{1}{2}B_{i_p},$$
$$\text{with } \Delta_{i_p} = B_{i_p}^2 - 4C_{i_p} = 0 \ (i_p \in \{i_1, i_2, \dots, i_l\}), \tag{6.93}$$
$$a^+_{\sum_{i=1}^{p-1} l_i+1} \neq \dots \neq a^+_{\sum_{i=1}^{p-1} l_i+l_p} \text{ or } a^-_{\sum_{i=1}^{p-1} l_i+1} \neq \dots \neq a^-_{\sum_{i=1}^{p-1} l_i+l_p}.$$

(b) The functional equilibrium of $x^* = a_{i_p}^{(s)} \in S_{i_p}$ with $g(x^*) = a_{i_p}$ for $(l_p > 1)$-repeated functional equilibriums switching is called an l_pth XX bifurcation of $(l_{p_1}\text{th XX}:l_{p_2}\text{th XX}:\dots:l_{p_\beta}\text{th XX})$ functional equilibrium switching at a point $\mathbf{p} = \mathbf{p}_1 \in \partial\Omega_{12}$, and the functional bifurcation condition is

$$a_{i_p} \equiv a_{\sum_{i=1}^{p-1} l_i+1} = \dots = a_{\sum_{i=1}^{p-1} l_i+l_p},$$
$$a^\pm_{\sum_{i=1}^{p-1} l_i+1} \neq \dots \neq a^\pm_{\sum_{i=1}^{p-1} l_i+l_p}, \tag{6.94}$$
$$l_p = \sum_{i=1}^{\beta} l_{p_i}, \quad p = 1, 2, \dots, n_1.$$

(c) The equilibrium of $x^* = a_{i_p}^{(s)} \in S_{i_p}$ with $g(x^*) = a_{i_p}$ for $(l_{p_1} \geq 2)$-repeated functional equilibriums appearance/vanishing and $(l_{p_2} \geq 2)$ repeated functional equilibriums switching of $(l_{p_{21}}\text{th XX}:l_{p_{22}}\text{th XX}:\dots:l_{p_{2\beta}}\text{th XX})$ is called an l_pth XX bifurcation of functional equilibrium at a point $\mathbf{p} = \mathbf{p}_1 \in \partial\Omega_{12}$, and the functional

bifurcation condition is

$$a_{i_p} \equiv a_{\sum_{i=1}^{p-1} l_i + 1} = \cdots = a_{\sum_{i=1}^{p-1} l_i + l_p}$$

$$\text{with } \Delta_{i_p} = B_{i_p}^2 - 4C_{i_p} = 0 \; (i_p \in \{i_1, i_2, \ldots, i_l\})$$

$$a^+_{\sum_{i=1}^{p-1} l_i + j_1} \neq \cdots \neq a^+_{\sum_{i=1}^{p-1} l_i + j_{p_1}} \text{ or } a^-_{\sum_{i=1}^{p_1-1} l_i + j_1} \neq \cdots \neq a^-_{\sum_{i=1}^{p_1-1} l_i + j_{p_1}},$$

$$\text{for } \{j_1, j_2, \ldots, j_{p_1}\} \subseteq \{1, 2, \ldots, l_p\},$$

$$a^{\pm}_{\sum_{i=1}^{p-1} l_i + k_1} \neq \cdots \neq a^{\pm}_{\sum_{i=1}^{p-1} l_i + k_{p_2}}$$

$$\text{for } \{k_1, k_2, \ldots, k_{p_2}\} \subseteq \{1, 2, \ldots, l_p\},$$

$$\text{with } l_{p_1} + l_{p_2} = l_p.$$

$$(6.95)$$

(iv) If

$$\Delta_i = B_i^2 - 4C_i > 0 \text{ for } i = l + 1, l + 2, \ldots, n, \qquad (6.96)$$

the $(2m + 1)$th-degree polynomial nonlinear system has $(2m + 1)$ equilibriums as

$$g(x^*) = b_1^{(i)} = -\frac{1}{2}(B_i + \sqrt{\Delta_i}),$$

$$g(x^*) = b_2^{(i)} = -\frac{1}{2}(B_i - \sqrt{\Delta_i})$$

$$\text{for } i = l + 1, l + 2, \ldots, n;$$

$$\{a_1, a_2, \ldots, a_{2n+1}\} = \text{sort}\{c_1, c_2, \ldots, c_{2l+1}, \underbrace{b_1^{(r+1)}, b_2^{(r+1)}}_{q_{r+1} \text{ sets}}, \ldots, \underbrace{b_1^{(n)}, b_2^{(n)}}_{q_n \text{ sets}}\}, \quad (6.97)$$

$$\text{with } a_\alpha < a_{\alpha+1}$$

$$S_\alpha = \{a_\alpha^{(s_\alpha)} | g(a_\alpha^{(s_\alpha)}) = a_\alpha, \; s_\alpha = 1, 2, \ldots, N_\alpha\} \cup \{\emptyset\},$$

$$\alpha = \{1, 2, \ldots, 2n + 1\}.$$

If

$$\{a_1, a_2, \ldots, a_{2n+1}\} = \text{sort}\{c_1, c_2, \ldots, c_{2l+1}, \underbrace{b_1^{(r+1)}, b_2^{(r+1)}}_{q_{r+1} \text{ sets}}, \ldots, \underbrace{b_1^{(n)}, b_2^{(n)}}_{q_n \text{ sets}}\},$$

$$a_{i_1} \equiv a_1 = \cdots = a_{l_1},$$

$$a_{i_2} \equiv a_{l_1+1} = \cdots = a_{l_1+l_2},$$

$$\vdots \qquad (6.98)$$

$$a_{i_n} \equiv a_{\sum_{i=1}^{n-1} l_i + 1} = \cdots = a_{\sum_{i=1}^{n-1} l_i + l_r} = a_{2n+1}$$

$$\text{with } \sum_{s=1}^{n} l_s = 2m + 1,$$

then the corresponding standard functional form is

$$\dot{x} = a_0 \prod_{s=1}^{r} [g(x) - a_{i_s}]^{l_s}.$$

(6.99)

The functional equilibrium separatrix flow is called an $(l_1 \text{th XX} : l_2 \text{th XX} : \ldots : l_r \text{th XX})$-functional flow. The functional equilibrium of $x^* = a_{i_p}^{(s)} \in S_{i_p}$ with $g(x^*) = a_{i_p}$ for l_p-repeated functional equilibriums switching is called an $l_p \text{th XX}$ switching bifurcation of $(l_{p_1} \text{th XX} : l_{p_2} \text{th XX} : \ldots : l_{p_\beta} \text{th XX})$ functional equilibrium at a point $\mathbf{p} = \mathbf{p}_1 \in \partial \Omega_{12}$, and the functional bifurcation condition is

$$a_{i_p} \equiv a_{\sum_{i=1}^{p-1} l_i + 1} = \cdots = a_{\sum_{i=1}^{p-1} l_i + l_p},$$

$$a_{\sum_{i=1}^{p-1} l_i + 1}^{\pm} \neq \cdots \neq a_{\sum_{i=1}^{p-1} l_i + l_p}^{\pm};$$

$$l_p = \sum_{i=1}^{\beta} l_{p_i}, \quad p = 1, 2, \ldots, r.$$

(6.100)

Author's Biography

ALBERT C. J. LUO

Professor Luo works at Southern Illinois University, Edwardsville. For over 30 years, Dr. Luo's contributions on nonlinear dynamical systems and mechanics lie in (i) the local singularity theory for discontinuous dynamical systems, (ii) dynamical systems synchronization, (iii) analytical solutions of periodic and chaotic motions in nonlinear dynamical systems, (iv) the theory for stochastic and resonant layer in nonlinear Hamiltonian systems, and (v) the full nonlinear theory for a deformable body. Such contributions have been scattered into over 20 monographs and over 300 peer-reviewed journal and conference papers. Dr. Luo served an editor for the journal *Communications in Nonlinear Science and Numerical Simulation*, book series on *Nonlinear Physical Science* (HEP), and *Nonlinear Systems and Complexity* (Springer). Dr. Luo was an editorial member for *IMeChE Part K Journal of Multibody Dynamics* and the *Journal of Vibration and Control*, and has organized over 30 international symposiums and conferences on Dynamics and Control.

Printed in the United States
by Baker & Taylor Publisher Services